INNOVATION

To my advisors, friends and sons:
Lucien and Doug

INNOVATION

THE ATTACKER'S
ADVANTAGE

Richard N. Foster

MACMILLAN
LONDON

First published in the United States of America 1986 by Summit Books,
a division of Simon & Schuster, Inc., New York

First published in the United Kingdom 1986 by
MACMILLAN LONDON LIMITED
4 Little Essex Street London WC2R 3LF
and Basingstoke

Associated companies in Auckland, Delhi, Dublin, Gaborone,
Hamburg, Harare, Hong Kong, Johannesburg, Kuala Lumpur,
Lagos, Manzini, Melbourne, Mexico City, Nairobi, New York,
Singapore and Tokyo

ISBN 0-333-43511-7

Printed and bound by the Garden City Press Limited
Letchworth, Herts SG6 1JS

CONTENTS

Acknowledgments

The research and consulting condensed in this book represents many years of productive collaboration with my clients and my colleagues in McKinsey & Company. McKinsey works with some of the largest and most successful companies in the United States, Japan and Europe. These companies have sophisticated managers who often must invent new ways to solve new problems. Many of them are concerned with technology. Both directly and through my partners I was able to work on these problems and in the process test the ideas in this book. That would not have been possible without their trust and confidence in me and my colleagues and their willingness to break new ground.

A few people at McKinsey deserve special mention. First and foremost is my mentor, teacher and friend, Fred Gluck. Fred and I have worked together on the problems of technology and strategy for over a decade. He has always been the most creative, insightful and yet rigorous thinker I have known. His constantly fresh approach to problems, no matter how many times addressed before, is inspiring. We have done so much together that it is no longer possible for me to know where my ideas stop and his begin. In many ways he is the co-author of this book.

Ron Daniel, the managing director of McKinsey, has over

the years always backed my interests and research. Mike Bulkin, manager of our New York office, has had more patience than anyone could reasonably expect when I was off in seclusion working on this book. Without their support and encouragement none of the activities that led to this book could have been possible.

My colleagues Ed Krubasik in Munich, Art Chivvis and Ali Hanna in New York, and Bill Lewis in Washington have been constant co-conspirators. They have made substantive contributions to the thoughts put forward here. Ed has applied them to a wide range of problems in the electronics and aerospace industries in Europe, Art to the problems of drug discovery, Ali to the telecommunications and aircraft business. Bill Lewis has led the charge in petroleum, chemicals and metals. Somu Sobramaniam, an economist and financial wizard, has made more contributions to understanding the economic implications of technological change than anyone I know. He has been aided and abetted by a number of others including Greg Summe from Atlanta, Pauline Walsh from Toronto, and Wolfgang Leitner from Munich. Diana Mackie has applied our thinking in many areas including medical technology, consumer nondurables and beverages. Tom Woodard has been a constant source of encouragement and support over the years by passing on ideas and examples that have proved to be useful.

Tom Wilson, who heads McKinsey's consumer practice, has pushed these ideas into the consumer arena, where I was not sure they applied. Bob Conrads, who heads McKinsey's electronics practice, has brought them to bear on problems in that fast-moving industry. Tom Steiner has done the same in banking. Steve Walleck has taken the lead in the manufacturing and robotics area.

Larry Linden deserves special attribution for his provocative suggestions both when he served on the White House science and technology staff and later when he joined McKinsey. Larry has played the lead role in joint activities we have had over the past few years with the Industrial Research Institute, the association of R&D VPs of this country's largest companies.

Ken Weisshaar and Charles Jones from the U.S., Tadaaki Chigusa and Partha Ghosh from Japan, Mickey Obermayer from Copenhagen, Jurgen Schrader from Munich, Roger Abravanel from Milan and George Norsig when he was in Mexico and now in Dallas, and Steve Schwarzwaelder in Pittsburgh have all made important contributions. So have Julien Phillips, Dick Cavanagh, Don Clifford, Ken Ohmae, Henry Strage, and Bob O'Block.

Since we began our efforts, a few of the collaborators in this effort have gone on to do other things. First is Russ Craig, who is now Vice President for Strategic Planning at Genrad in Boston. Russ worked out many of the initial examples of the S-curve theory that have stood the test of time. Bob Waterman certainly influenced my thinking about corporate renewal, a subject we discussed many times. Roberto Buaron, who has now entered the venture capital world, brought insight to the relationship between technological change and competitive position. Eddie Miller, formerly with McKinsey in London but now also a venture capitalist in New York, also applied the ideas to the electronics industry.

The list is longer—too long to provide individual recognition—but I wish to thank all my colleagues at McKinsey who through their study, work and comments have pushed this thinking forward.

No work like this can be done without a lot of research support, and for this I thank Matt Palmieri and Gail Mueller. They have proved tireless in seeking out facts, no matter how arcane. (What was the production of watches in Switzerland from 1750–1800?) They have also been skillful diplomats when they found data that did not agree with my ideas.

In the process of doing this book we also made a video tape for internal training. This also pushed our thinking. Tim Lynch wrote the script and turned our clumsy prose into effective English. He is one of the best writers on technology management I know. Judy Owens Bergsma was the producer. Shé did a wonderful job with an unruly bunch.

Needless to say, this book also required enormous secretarial assistance that was provided above and beyond any reasonable requests by Marilena Christoforou, my secretary, who was also carrying a full workload as well as evening school classes. Never during the endless drafting and redrafting was she anything but enthusiastic. Bill Price was also terrific pulling together and editing the final manuscript at McKinsey.

To my publisher and editor, James H. Silberman and Arthur H. Samuelson, go my thanks not only for a job well done but for the vision to see the book before there was any evidence that it would be a book, and for their patience and guidance through the various drafts. They have taught me a lot about how to present ideas.

Paul Gibson, Vice President at Hill & Knowlton who for ten years wrote for *Forbes* magazine, deserves special thanks. Paul somehow managed to distill the 1,000-page volume I wrote while at Martha's Vineyard two summers ago to a workable draft. His work and toughmindedness helped me take an important step toward publication.

Most specially, thanks go to my friend and inspiror Bill Matassoni, Vice President of Communications at McKinsey. Over the three to four years when this book was forming and taking shape, Bill provided constant good advice. More important, at the end of the whole process, when a special push was needed, Bill turned his full efforts to *Innovation: The Attacker's Advantage* and helped me write its final draft. Without him there would have been no book.

Foreword

_____BY SIR HENRY CHILVER

*F*or more than a century now, industrial technology has played a central role in the development of the world economy and world society. So central has its role been that all countries of the world—at some stage or other—have aspired to the successful exploitation of industrial technology to provide the main engine of economic and social change.

The most "advanced" countries economically are those which have the most outstanding success in exploitation of industrial technology. At the same time, the introduction of new technologies and new ideas into less developed countries has been seen universally to provide the main forces for progress in those less-advanced countries.

National and international agencies, dedicated to the wider dissemination of new technologies and new ideas, work in considerable numbers around the world. Over the decades, national governments have increasingly based their policies on the successful introduction to their countries of new ideas, particularly in the technologies.

But the most important "components" of industrial technology are the many organizations which exploit new ideas and trade them across the world. In the free world these are the "companies" which abound in both national and international scenes. The most successful of these companies are expert in developing—in timely ways—innovative products or services.

The progress of modern industrial technology is in fact dependent wholly on the abilities of the many companies around the world to innovate successfully. There have been a number of different schools of thought on how and where this innovation occurs. At one extreme it is argued that innovation is achieved within one organization and exploited by others; it is claimed that successful companies exploit the innovations made by other, less successful companies. At another extreme, it is argued that innovation should be nurtured entirely within any organization, without recourse to external ideas.

In this fascinating and thoroughly stimulating book, Richard Foster comes out firmly in favour of the theory that the most successful companies are also masters of the art of innovation. Their success depends in fact utterly on their ability to develop new products and to market and sell these. Moreover, all these must be achieved in very timely fashion.

The evidence for this in the world of industrial technology is overwhelming. In tracing the basis of this whole philosophy, Richard Foster shows the historical evolution of company products and the ways in which the fortunes of companies depend on their abilities to break away from dependence on mature products and to innovate successfully in the introduction of new products.

In such a situation, the *attacker's* advantage is crucial. The attack must be carefully planned and timed. Strategies of defence suffer from serious weaknesses. In defence, companies will always be put at increasing disadvantages, which will eventually bring them down.

Many of those who work at the "coal-face" of development of industrial technology will recognize in this book the basis of the process of development of which they have been a part. Those organizations which choose not to attack will encounter serious problems of development. As Richard Foster points out, the best form of defence is in fact attack.

This outstanding book is written with great clarity. The case studies are presented starkly and effectively. This style makes the book especially attractive to those senior people in industry who have paused in their day-to-day work and thought more deeply about the broader nature of the development process which is at work in modern commercial industrial technology.

This book deserves to be widely read. It should be taken seriously by both scholars and industrialists in any country, and in any company which purports to understand the fundamental forces at work in industrial technology. We are all indebted to Richard Foster for the graphic way in which he has presented this compelling picture of innovation.

Cranfield, 1986

*T*his book began years ago, about the time I finished my Ph.D. thesis on the binary flow of nonabsorbable gases through porous media. Later I wondered why I did that particular thesis. Why had I been given that technical problem to work on and not some other one?

At Union Carbide, where my principal concern was marketing, I assumed that most people had answers to these and related questions. That they believed that there was a connection between R&D and profits. But many people did not. They saw R&D as overhead, an expense to be minimized. At NASA, where I next worked as a consultant, developing commercial applications for aerospace technology, I continued to ask the same general question in different ways. How does one establish a research program and set its priorities? What is the connection between technology development and corporate success? Indeed, is there any connection? What are corporations getting for their R&D expenditures, which were $45 billion in the U.S. alone in 1984.

Even at McKinsey, where my colleagues and I had the chance to work with large, well-managed companies around the world, it wasn't clear that our faith in the importance of research and development was well founded. We felt something was missing in the way companies generally managed their R&D, but we weren't sure what it was.

Several of us met frequently in the late 1970s to trade ideas about business strategy. We would gather at regular intervals in Switzerland, and it was there, after rereading Thomas Kuhn's *The Structure of Scientific Revolutions,** that I finally began to understand the pace of technological progress—how a technology advances slowly at first, then accelerates, and then inevitably declines. At the end of our session, I made some notes and presented them to the group. We all felt that we were beginning to get some answers to the questions about R&D and its value to companies.

Not long afterward, I discussed our emerging ideas with some people at Exxon. They were supportive but said that something was missing—something important. It was necessary, they said, to understand the pace of technological progress, but it was also important to discuss profits. They were right. Recognizing that, we were on the way to putting together a complete picture of the impact of R&D on corporate success. The progress we made forced us to take a different cut at innovation than others who had studied it.

To most observers, innovation is a solitary process that requires creativity and genius, perhaps even greatness. It can't, in their view, be managed or predicted, just hoped for and perhaps facilitated. But for me innovation was and still is more than that. It was a battle in the marketplace between innovators or attackers trying to make money by changing the order of things, and defenders protecting their existing cash flows. That is the perspective of *Innovation: The Attacker's Advantage.* This point of view enabled me and my colleagues to explore the economics of innovation—some-

* Kuhn, Thomas S., *The Structure of Scientific Revolutions,* Chicago: University of Chicago Press, 1970.

thing that the "great man" school of innovation had largely ignored. We had to look at corporate success and failure over long periods of time (e.g., 20–25 years) to see patterns, but patterns did emerge. We began to see not only patterns of success and failure but the principles that caused events to unfold as they did. There was indeed a structure and predictability about innovation. These patterns suggested that in most cases it is companies with new ideas and approaches, not entrenched large ones, that collectively have the advantage—the attacker's advantage. These ideas also helped us determine when the attack is most likely to succeed and when it is most likely to fail. Many early attacks fail—as do those that are delayed too long. So these insights were valuable.

Since then, through work with our clients, this basic premise has been refined and extended. Most important, we now have a better understanding of the economics of innovation—when the impact of new products is first likely to be felt, what will happen to prices and profits for the attackers and defenders, how long the battle will last. Working on an article for *Business Week* with editor Alice Priest, I came to grips with the last key idea of this book—the need to attack yourself at the same time you are making a good defense.

In the end, *Innovation: The Attacker's Advantage* came to be about companies that have more up years than their competitors because they recognize that they must be close to ruthless in cannibalizing their current products and processes just when they are most lucrative, and begin the search again, over and over. It is about the inexorable and yet stealthy challenge of new technology and the economics of substitution which force companies to behave like the mythical phoenix, a bird that periodically crashed to earth in order to rejuvenate itself. It is thus not a book about process—seven steps to new products—but instead about a point of view about business in which change, success and failure seem to come more and more rapidly.

Henry Ford, long ago, understood the message of this book. In *My Life and Work*,* he wrote, "If to petrify is success, all one has to do is to humor the lazy side of the mind; but if to grow is success, then one must wake up anew every morning and keep awake all day. I saw great businesses become but the ghost of a name because someone thought they could be managed just as they were always managed, and though the management may have been most excellent in its day, its excellence consisted in its alertness to its day, and not in slavish following of its yesterdays. Life, as I see it, is not a location, but a journey. Even the man who most feels himself 'settled' is not settled—he is probably sagging back. Everything is in flux, and was meant to be. Life flows. We may live at the same number of the street, but it is never the same man who lives there." And later in the same book: "It could almost be written down as a formula that when a man begins to think that he at last has found his method, he had better begin a most searching examination of himself to see whether some part of his brain has not gone to sleep."

This book is that examination. It recognizes that innovation is born from individual greatness but contends that it lives in the province of the marketplace—that it is a repeatable economic event. The underpinnings of this event are best depicted by using a chart in the shape of an S-curve. With it and the analyses it forces, we can answer questions such as how much change is possible, when it will occur and how much it will cost. To get the answers requires standard economic analyses but also information about technology that most companies do not obtain or use. Without it they cannot

* Ford, Henry, with Crowther, Samuel, *My Life and Work*, Garden City, NY: Doubleday & Co., 1922.

renew themselves because they lack the conviction that they must. They believe the past is prologue. They are not convinced of how quickly the world will make their products and people less valuable. They don't believe that the attackers really do have the advantage. Perhaps now they will.

ONE

WHY LEADERS BECOME LOSERS

To cherish traditions, old buildings, ancient cultures and graceful lifestyles is a worthy thing—but in the world of technology to cling to outmoded methods of manufacture, old product lines, old markets, or old attitudes among management and workers is a prescription for suicide.

—Sir Ieuan Maddock
New Scientist, 1982

Toward dawn on Friday, December 13, 1907, the sailing ship *Thomas W. Lawson* sank off the Scilly Isles in the English Channel. All but the captain and one crew member were lost. This would have been just another Friday the thirteenth shipwreck, but the *Thomas Lawson* was special. It was a beautiful, huge ship with seven masts (Exhibit 1). It had been designed to compete against the new steam-powered vessels that had increasingly taken cargo business away from sailing ships. Constructed by the Fall River Ship and Engine Building Company, the *Thomas Lawson* could travel at 22 knots if the winds were brisk. But to gain swiftness her designer had had to sacrifice maneuverability. She was unwieldy and difficult to handle. In fact, she was so unstable she capsized while at anchor during a severe gale. According to one account, she was found in the morning looking "not unlike the back of a whale . . . the vast hull on its side washed by the seas." No attempt was ever made to design a faster cargo-carrying sailing ship. The builders and their employees sought other things to do. The age of commercial sail ended with the *Thomas Lawson,* and steamships began to rule the seas.

In May 1971, National Cash Register of Dayton, Ohio, stunned its workers, managers and investors by announcing

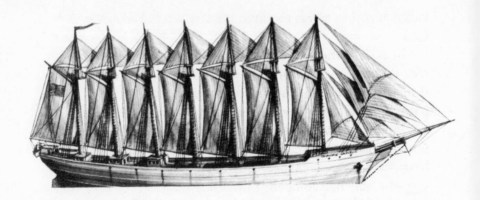

1   The *Thomas W. Lawson,* 1902 to 1907.
The *Lawson*'s seven masts crowded as much sail above her decks as the
limits of space and windflow would allow.
Source: Angelucci, Enzo, and Cucari, Attilio, *Ships,* New York: McGraw-Hill, 1975.

that $140 million worth of newly designed cash registers
were impossible to sell and would be written off. In the
months that followed, thousands of workers were laid off
and the CEO was fired. The stock price fell from 45 to 14
over the next four years. The problem? The machines used
electromechanical parts and could not compete with new,
cheaper to make, and easier to use electronic machines.

In 1947, Procter & Gamble introduced Tide, the first
synthetic laundry detergent. It was superior to the conventional natural detergents because it contained phosphate
"builders" which improved its cleaning power. At that time
P&G's major competitor in detergents was Lever Brothers.
But Tide changed all that. Its sales took off, leaving Lever far
behind, unable to match P&G's technical achievements.
Lever eventually responded with its own synthetic product

called Surf, but it was too little too late. P&G had stolen the lead.

In each of these cases and many more like them, companies that were leaders in their field saw their fortunes suddenly disappear. Do leading companies in fact not have the natural advantages they are supposed to or are their natural advantages outweighed by other inherent disadvantages? I think the latter is the case, and that these disadvantages result from technological change. Technological change is the reason why only one company out of three manages to cover its cost of raising money most of the time. Why most companies manage to achieve what Bob Waterman and Tom Peters have defined as excellent financial performance in only one year out of twenty, and then immediately drop back into the great middle ground of average financial performance. Why even the best companies, by anyone's definition of excellence, retain their superior competitive performance for only three to four years. Why only one manufacturing company in the United States during the last twenty years, Xerox, has been able to sustain a position of financial leadership* in its industry for ten years.

_____ UNDERSTANDING THE DYNAMICS OF COMPETITION

The roots of this failure lie in the assumptions behind the key decisions that all companies have to make. Most of the managers of companies that enjoy transitory success assume that tomorrow will be more or less like today. That significant

---

* Financial leadership means being ranked in the top third of industry in return on equity and sales growth for at least half of the 18-year period from 1965 to 1983.

change is unlikely, is unpredictable, and in any case will come slowly. They have thus focused their efforts on making their operations ever more cost effective. While valuing innovation and espousing the latest theories on entrepreneurship, they still believe it is a highly personalized process that cannot be managed or planned to any significant extent. They believe that innovation is risky, more risky than defending their present business.

But companies like IBM, Hewlett-Packard, Procter & Gamble, Johnson & Johnson, United Technologies, Harris, and Corning have all made the opposite assumptions. Their managers have assumed that the day after tomorrow will not be like today. They have assumed that when change comes it will come swiftly. They believe that there are certain patterns of change which are predictable and subject to analysis. They have focused more on being in the right technologies at the right time, being able to protect their positions, and having the best people rather than on becoming ever more efficient in their current lines of business. They believe that innovation is inevitable and manageable. They believe that managing innovation is the key to sustaining high levels of performance for their shareholders. They assume that the innovators, the attackers, will ultimately have the advantage, and they seek to be among those attackers, while not relinquishing the benefits of the present business which they actively defend. They know they will face problems and go through hard times, but they are prepared to weather them. They assume that as risky as innovation is, not innovating is even riskier.

The beliefs of successful companies are not accidental; they are based on an understanding of the dynamics of competition. To understand these dynamics, which manifest themselves every day in the business press, we need to understand three ideas: the S-curve, the attacker's advantage, and discontinuities.

_____ *THE S-CURVE*

The S-curve is a graph of the relationship between the effort put into improving a product or process and the results one gets back for that investment. It's called the S-curve because when the results are plotted, what usually appears is a sinuous line shaped like an S, but pulled to the right at the top and pulled to the left at the bottom (Exhibit 2).

Initially, as funds are put into developing a new product or process, progress is very slow. Then all hell breaks loose as the key knowledge necessary to make advances is put in place. Finally, as more dollars are put into the development

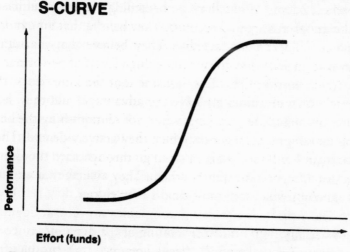

S-CURVE

Performance

Effort (funds)

2   The S-Curve.
The infancy, explosion, then gradual maturation of technological progress.

of a product or process, it becomes more and more difficult and expensive to make technical progress. Ships don't sail much faster, cash registers don't work much better, and clothes don't get much cleaner. And that is because of limits at the top of the S-curve.

_____ LIMITS: A NEW WAY TO THINK ABOUT TECHNOLOGY

Limits are fundamental to both our personal and our business lives. In everything we do or make we are governed by limits. We cannot go beyond them, so when we approach them we must change or not progress anymore. We all implicitly understand that. In July 1985, Sergei Bubka of the Soviet Union broke the world pole vault record, clearing 6 meters (19 feet 8¼ inches) during the Paris International Track and Field meet. Asked if he ever expected to reach 7 meters (22 feet 11¾ inches), he replied, "No, there will have to be another *technical revolution* before that height can be reached." There's just so high you can jump using a bending fiberglass pole.

In the world of business, limits determine which technologies, which machines and which processes are about to become obsolete. They are the reason why products eventually stop making money for companies. Management's ability to recognize limits is crucial to determining whether they succeed or fail, because limits are the best clue they have for recognizing when they will need to develop a new technology.

By technology I mean several things. In some cases it's a specific process—say a chemical process—that produces a specific product. In this case it's hard to distinguish the product from the technology. More broadly, technology can mean a manufacturing process, say continuous casting of steel versus the open-hearth method. Here the technology is distinct from the product. The cash management account (CMA) is another example of a distinct process and product. New information processing technology made the CMA possible. We can think of technology even more broadly as the general way a company does business or attempts a task—the production line versus batch processing or the sidelong high jump technique versus the backward-first Fosbury flop.

The point is this: technology even variously defined always has a limit—either the limit of a particular technology, for example, the ultimate density of devices we can get onto a silicon chip, or a succession of limits of several technologies that together make up the larger technology or product or way of doing business.

It's easy to see how these limits will affect performance and sales of a product when the technology and product are closely associated. It's not so easy to see the importance of limits when you are dealing with something like air travel, which results from thousands of technologies. But usually there are one, two or several technologies that are crucial to a product or its production (the semiconductor chip in a computer or the pole in the vaulter's hands), and these are the technologies with which managers, inventors and all of us ought to be concerned.

All of us know about limits, but often companies do not recognize or act on them. This lesson was brought home to me a couple of years ago when I was visiting a paper mill in Alabama. This particular company had erected a new mill alongside an old one. Our tour began in the old plant. As we strolled through the control room I could see engineers

F:I-2

watching their production statistics being automatically tabulated on paper charts and computer forms. I asked about the possibility of electronics replacing paper, say by having TV monitors replace computer forms in office applications. My host graciously but firmly told me this would never happen. Paper was too much a part of our lives. We needed to feel it and touch it. Without it we could lose the feeling of security, of possession that paper confers. Then we moved on into the control room of the new plant. There was no paper in sight, only banks of electronic screens! I felt as though I were talking to the captain of the *Thomas Lawson*. This manager didn't understand that the limits of printing on paper as a technology for conveying information were not far away and that electronic technology would soon be able to convey information more effectively and cheaply.

If you are at the limit, no matter how hard you try you cannot make progress. As you approach limits, the cost of making progress accelerates dramatically. Therefore, knowing the limit is crucial for a company if it is to anticipate change or at least stop pouring money into something that can't be improved. The problem for most companies is that they never know their limits. They do not systematically seek the one beacon in the night storm that will tell them just how far they can improve their products and processes.

That's not always the case. It was an understanding of the limits of its current technologies that persuaded IBM to develop a new semiconductor packaging technology for the 4300 and 308X series of computers that they introduced in the late 70s; persuaded Bell Labs along with Corning to be the first to develop fiber optic cables for telecommunication in the 60s; and persuaded Sir James Black to reject the conventional screening approach for new drugs to come up with Tagamet for ulcers, which then put Smith Kline into a leadership position. As T. R. Reid described in *The Chip*,* it was an understanding of the "tyranny of numbers," the limits of connecting wires, that convinced Noyce and Kilby to develop a new process and product that eventually became the semiconductor chips that permeate so much of our lives today. At the time it was hard for outsiders to see why these companies and men abandoned their past successes. But the outsiders didn't understand limits.

For those who don't understand limits and the S-curve, change comes as a surprise, catching them on their blindside. It happens so often and predictably that I've often thought of calling the S-curve the "blindside curve." But this would emphasize the negative too much. The S-curve has a positive side too. Companies can and do use it as the basis for successful attacks. Indeed, we could call it the "attacker's curve" too. So let's leave it as the S-curve, a literal name for its shape.

THE ATTACKER'S ADVANTAGE

For the S-curve to have practical significance there must be technological change in the wind. That is, one competitor must be nearing its limits, while others, perhaps less experienced, are exploring alternative technologies with higher limits. But this is almost always the case. I call the periods of change from one group of products or processes to another, technological discontinuities. There is a break between the S-curves and a new one begins to form. Not from the same knowledge that underlays the old one but from an entirely

* Reid, T. R., *The Chip: The Microelectronics Revolution and the Men Who Made It*, New York: Simon & Schuster, 1985.

new and different knowledge base. For example, the switch from vacuum tubes to semiconductors, the switch from propeller-driven planes to jets, the switch from natural to synthetic detergents or fibers, the shift from cloth to paper diapers, the switch from records to tapes to compact discs, the switch from carbonated cola drinks to carbonated juice drinks, and even the shift from conventional tennis racquets to the Prince racquet with its enlarged "sweet spot." These are all technological discontinuities. And they have all unseated industry leaders.

Technological discontinuities have been and will be arriving with increasing frequency. The scientific knowledge that underpins our products and processes is multiplying by leaps and bounds in fields as diverse as quantum physics, surface chemistry, cell biology, mathematics and the structure of knowledge itself. Furthermore, every day we learn more about the process of innovation—how it works and how it can be made to work better. These two developments aren't new, but never before have they interacted in such a way to produce the explosion of knowledge and change that is taking place today. Thus, it's my feeling that as much as 80 percent of all manufacturing industries and a large portion of all service industries will see major technological changes before the year 2000. We are living in an age of discontinuity, and in an age in which the risk to industry leaders has never been greater.

The results of a discontinuity are almost always brutal for the defender. The failure to recognize the limits of electromechanical cash registers at NCR cost thousands of workers and executives their jobs. It cost NCR's investors millions of dollars as well. For Unilever it meant months spent scrambling to produce their own synthetic detergents all the while losing the lead to P&G. For our paper maker in Alabama it may mean underutilized plant capacity and depressed prices if electronics reduces the need for paper.

My thesis is not only that technological discontinuities will come with increasing frequency as we move toward A.D. 2000, but that during these discontinuities the attackers will have the advantage over the defenders.

As limits are reached, it becomes increasingly expensive to make progress. At the same time, the possibility of new approaches often emerges—new possibilities that frequently depend on skills not well developed in leader companies. As these attacks are launched, they are often unnoticed by the leader, hidden from view by conventional economic analysis. When the youthful attacker is strong he is quite prepared for battle by virtue of success and training in market niches. The defender, lulled by the security of strong economic performance for a long time and by conventional management wisdom that encourages him to stay his course, and buoyed by faith in evolutionary change, finds it's too late to respond. The final battle is swift and the leader loses. His attempts to defend his employees and shareholders fail. Doomed by doing too little, too late.

In order to overcome the attacker's advantage, defending companies must understand the S-curve and limits, because they will tell management when an attack can occur and what its consequences might be. They will thus help defenders anticipate and deal with their challengers.

_____ A NEW PARADIGM

This is not a theory. Companies have used these insights either implicitly or explicitly to get the jump on competition. For example, Procter & Gamble not only exploited synthetic detergents but also exploited opportunities in paper process-ing to come up with Pampers. Pampers now accounts for 35 percent of disposable diaper sales and more than 20 percent of P&G's profits. P&G is trying to do it again with their new process for making orange juice. Michelin captured 11 per-cent of the U.S. tire market when it introduced radials, which produced longer tire life. Citibank put its competitors on the defense when it introduced its automated tellers. Sony and other companies may eventually capture the recorded music market with the introduction of the compact disc that pro-vides much more realistic sound than audio tape. Smith Kline ensured its place as the number-one–ranked pharmaceutical company in terms of earnings and earnings growth when it introduced Tagamet for ulcer treatment, again a product based on a new technology. Johnson & Johnson did it with Tylenol. The Japanese gained an advantage over the Swiss with digital watches. GE did it to its competitors in jet en-gines with the high by-pass fan jet. Texas Instruments did it to Westinghouse, Sylvania and several other established com-panies when it captured the lead in solid-state electronics. Harris Intertype Corporation saved itself back in the mid-60s by switching from the maturing electromechanical type-setting technology to electronics. Xerox did it to the carbon paper makers during the 60s by developing its inherently more flexible and cheaper copying and duplicating process. U.S. Surgical did it in surgery by exploiting the potential of staples to close wounds, a procedure developed by the Rus-sians. IBM took the lead from Smith Corona in the office by developing electric typewriters, which have subsequently be-come computer-based word processors. In each case an emerging technological opportunity, juxtaposed with a ma-turing but still improvable traditional technology, provided an opportunity for a new competitor to grab leadership from an existing one.

If you look at the business landscape with this new paradigm and perspective of S-curves, limits and the attacker's advantage, you will see other marketplace battles shaping up. The makers of electronic cameras that put images on magnetic media could challenge the now dominant technology of chemical processing of pictures. Fotomats would be a thing of the past. Soft drinks based on juices developed with the aid of biotechnology may challenge colas. Specially bred plants resistant to insects and other pests will change the need for herbicides and insecticides now made by chemical companies. Parallel processing computers will challenge conventional serial processing computers and their makers. Indeed, makers of light-based computers may challenge makers of electron-based computers like IBM and DEC; and a new material called gallium arsenide may take a hunk of the silicon semiconductor market from its makers. Magnetic resonance imaging technologies will replace the need for computer aided tomography (CAT scans) in medical diagnosis. Monoclonal antibodies may turn out to be a safe and more effective substitute in cancer therapy for radiation and all its ill effects. New companies may exploit the potential of biotechnology to produce drugs that do not act directly, but rather stimulate the body's production of "natural" drugs which have no side effects. Home banking based on electronics may continue to replace the need for branch expansion at the major money center banks. The list can go on and on. The questions are which of these changes will occur and when. I don't have the answers, but it's possible to get them by doing the right analysis with S-curves and understanding limits.

In short, these ideas will give the reader a new understanding of competition. Needless to say the S-curves, limits analyses and other techniques described in this book can be misused. Mistakes can be made, even by scientists, about limits—particularly the limits of a competitor's technologies. Even if the limits are clearly defined, the breakthrough idea about how to reach them may be missing. Internal processes may be slower and more cumbersome than estimated, causing big increases in the cost of pushing a technology to its limit. These problems may lead to incorrect diagnosis of what the future holds and misguided actions. But if these mistakes are avoided, and they can be avoided most of the time, the S-curve will provide a solid base for thinking about what will happen in the future, and doing what is necessary to capitalize on opportunities.

Despite the fact that S-curves, limits analyses, and the attacker's advantage are concepts that are not simple to apply, they help to explain a number of ideas that often go against conventional management wisdom. Why leaders lose. Why there are no static advantages in business, and why individual products lose a competitive position faster than we expect.

Why small competitors often get the drop on large competitors. And why corporate leaders who attempt to control the pace of innovation in their industry almost always fail.

Consider the advice "stick to your knitting." Bob Waterman, who co-authored *In Search of Excellence*, wrote there that a corporation's success is based on the unique set of competitive skills it has honed over the years. Thus, he argued, it is best for it to stay close to home and not move into new fields. But what about when a company is near the limits of its ability to improve its products as NCR was in 1971? What should it do then? Curtiss-Wright tried staying with propeller aircraft long after the introduction of jets and is now just a shadow of its long-standing competitors Lockheed and McDonnell Douglas. Likewise, Addressograph-Multigraph, the maker of office products, stuck with mechanical machines that couldn't keep pace with electronic counterparts. In short, when it comes to technology the best strategy may be to do the unfamiliar. Move away from familiar areas into less familiar ones, as Harris, Corning and Gould have done. A frightening prospect personally and professionally.

THE COURAGE TO CHANGE

The S-curve, limits and attacker's advantages are at the heart of these problems and they also provide the key to solving them. For example, there are people, call them limitists, who have an unusual ability to recognize limits and ways around them. They ought to be hired or promoted. There are others who can spot ways to circumvent limits by switching to new approaches. They are essential too. Imaginary products need to be designed to understand when a competitive threat is likely to become a reality. Hybrid products that seem to be messy assemblages of old and new technologies (like steamships with sails) can sometimes be essential for competitive success. Companies can set up separate divisions to produce new technologies and products to compete with old ones. S-curves can be sketched and used to anticipate trouble.

None of this is easy. And it won't happen unless the chief executive replaces his search for efficiency with a quest for competitiveness. Indeed it is the chief executive who has the major role to play in making sure his company rides the wave of technology that continues to hit us. He need not be a scientist, but he must be someone who understands how science and innovation develop, someone with the conviction to insist that the company abandon its technology and skill base when everything in classic economic terms is going well, someone with a thick skin to endure the criticism that will come when the first steps toward new products and processes inevitably go astray or prove disappointing.

Top management has to develop a language and a facility for talking about and directing technology. We don't hear about technology in the boardroom, except for some progress reports that we pretend to understand or criticize, because we don't have a language or conceptual framework for managing technology. There have been several theories proposed to help management "link" technology with the market by charting a company's strength in a particular technology against its market potential, but the link is often only visual and superficial. Indeed there is no understanding of the linkage. That is what this book hopes to do—establish the linkage and the language management needs in an era driven by technological discontinuity and international competition.

Most top executives understand, I think, that technological change is relevant to them and that it is useless and misleading to label their business as high-tech or low-tech. What they don't have is a picture of the engines of the process by which technology is transformed into competitive advantage and how they can thus get their hands on the throttle.

Because they lack the language and the right questions, they can't answer the big questions: How much trouble is my company really in? Does this new product or process represent a real threat? What is the long-term verdict? In the middle of the competitive battle when there is smoke on the field and people and products are falling over, that is what they need to know.

Understanding S-curves, the way they unfold and what will limit them, is key to diagnosing just how fatal is the trouble you are in. And how big an opportunity might be out there. These curves need not be drawn in retrospect. They can be sketched now. Precision isn't as important as point of view. It's enough to know the rough shape of a technology's approach in order to make good judgments. If change occurs at the time learning starts to slow, wrote Phillip Moffitt recently in *Esquire*, ". . . then there is a chance to avoid the dramatic deterioration. If we call this the 'observation point,' when you can see the past and the future, then there is time to reconsider what one is doing."*

That's what has been missing. Perhaps that's why so many executives have lacked the will to manage technology and have retreated to the position that technological developments are unpredictable and unmanageable.

My observation and that of my McKinsey colleagues is that even when top managers understand what is necessary to stay ahead, only a handful have the conviction and discipline to act on that understanding. Only a few have the will to change and have led their companies through technological discontinuities. In fact they have made them happen. The question is how many of us can follow their lead and avoid the fate of the *Thomas Lawson*.

* Moffitt, Phillip, "The Dark Side of Excellence," *Esquire*, December 1985, pp. 43–44.

TWO

THE AGE OF
DISCONTINUITY

The new system was not built by politi-
cians or economists. It was built by technology. In some respects the new world
financial system is the accidental by-
product of communication satellites and
of engineers learning how to use the
electromagnetic spectrum up to 300
gigahertz.

—Walter Wriston, former chairman of Citicorp
1985

Jim Utaski, the President of Johnson & Johnson's Baby Products company, is a broadly experienced business-man. He has worked in marketing in the United States and headed J&J operations in Brazil. Jim says that learning to cope with 100-percent-a-year inflation in Brazil is much eas-ier than learning how to cope with technological change. "Once you accepted the fact that inflation was there and going to stay, then you could deal with it. You knew what you had to do with inventories, with prices, with employee benefits. But with technological change it's always a new ball game. It's always new rules. You never know what to expect next, or who to expect it from. Plus there's the whole lan-guage question. Learning Portuguese was a lot simpler than learning a lot of the technical jargon."

These words might not be surprising if they came from the chairman of IBM or GE or Merck. But coming from the head of a company that makes talcum powder and No More Tears shampoo, they show how technology affects and chal-lenges every company, not just those labeled "high tech." Nevertheless, there is a point of view among businessmen that goes something like this: "Well, that's fine, but in reality discontinuities are few and far between. What most of life is about is continuity. Managing that is what I'm concerned about, not trying to manage something that may affect me only once in my life, if at all."

If only this were so. But the title of Peter Drucker's recent book, *The Age of Discontinuity,* does in fact describe the commercial era in which we live. Discontinuities do occur more frequently than most of us realize. And if anything, their frequency is on the increase. It's hard to find an industry where they're not happening or are looming on the horizon. And their ramifications can be enormous, creating an almost unending chain of commercial events that spell success for some and losses for others.

Take a trip to your supermarket freezer and look at the discontinuities happening in a product like orange juice. The technology of reconstituting orange juice, "squeezing" it into a can, was developed first by Beatrice Companies' Tropicana. Then came freeze storage by Coca-Cola's Minute Maid. Now cryogenic extraction is being commercialized by Citrus Hill, a division of Procter & Gamble. It's too early to tell yet if cryogenics will replace freeze storage, but it's always important to watch P&G, given its track record of creating new products with new technologies (e.g., Pampers and synthetic detergents).

A big change has taken place in the YMCAs, gyms and fitness centers because of technology. Free weights and pulley systems have been replaced by Nautilus equipment. Its inventor, Arthur Jones, calls it a "thinking man's barbell." As described by *Ultrasport* magazine, "the various Nautilus machines or 'stations' transform the one-directional resistance of a barbell (downward only, in response to gravity) into a rotational, multi-directional resistance more appropriate to the human body. Second, by using 'cams,' or odd-shaped, ellipsoid moving parts, the machines vary the resistance in a way that is tailored to the body's capability; more resistance where a muscle is strongest, less where it is weakest."* As a result, working out on Nautilus is much faster and more effective.

Banking, trade and the service industries are already changing dramatically because of computers and communications. Consider how the banks have used electronics to replace tellers with automatic machines. The magnetic strips on bank cards may soon be replaced by computer chips. As a result the card would become a record of your account rather than simply provide access to it, giving you instant updates and access to cash at department stores. Electronics is already eliminating paper in the back offices of banks for some applications and will eventually reduce costs and increase responsiveness. Electronics will also allow small bank branches to be created on street corners, and perhaps eliminate the need for larger branches, thus changing our whole banking structure. A group of entrepreneurs who wanted to start a new bank today using new technologies might not only be competitive with traditional banks but would probably have a substantial cost advantage. Citibank already ranks among the largest corporate spenders on R&D, and under new Chairman John Reed a technical background may be almost as important for career advancement as knowing how to handle financial transactions. Reed advanced to the top via consumer banking but holds a degree in physical metallurgy from MIT. His replacement in consumer banking is Richard Hoffman, who holds a doctorate in molecular biophysics from Yale. Hardly traditional bankers.

The Office of Technological Assessment expects annual sales of chemicals and drugs produced by gene splitting (biotechnology) to top the $15 billion mark before the turn of

* Roberts, David, "Citizen Jones," *Ultrasport*, September/October 1985, pp. 49–55.

the century. Biotechnology will affect many sectors. It will, for example, make over the once-stodgy seed business. Thanks to a rash of acquisitions, the traditional family-run concerns are giving way to big-league operators like Monsanto, Upjohn and Librizol. The driving force behind the acquisitions is a growing feeling among scientists that we are on the verge of a second "Green Revolution." Biotechnology will allow us to create hardier and more vigorous crops by introducing new genes directly into the plants. Scientists are developing plant varieties that can find their own nitrogen and are resistant to drought or high temperatures.

Molecular engineering may not get as much attention as genetic engineering, but its ramifications will be equally awesome. Some people refer to material scientists as molecule managers because they can move atoms around inside individual molecules to create entirely new forms of matter. Companies such as Hercules, Du Pont, Celanese and General Electric are giving us an impressive array of new materials that outperform traditional materials. Their work is the result of many patient years spent in the laboratories, and it comes just when it is needed. The aerospace industry is now using advanced composites in products like airplane wings, which have to withstand "high-fatigue" environments.

This is not good news for the makers of old-line metals like steel and aluminum. As one magazine put it, "No bugles will blow to announce the New Materials Age. But every day the products of molecular engineering are taking markets away from the traditional materials of the Industrial Revolution. For the manufacturer this means a trade-off: better performance for the headaches of switching to an unfamiliar process; lighter-weight and more energy efficiency in exchange for huge capital investments. But this, of course, is

THE AGE OF DISCONTINUITY

the essence of entrepreneurial opportunity: to become a beneficiary of change rather than a victim of change."*

There is tremendous potential here for substitution and replacement. We've already seen it in the packaging field with composite materials. We will undoubtedly see it in the chemical field as well. Polymers have been developed that can withstand 1000° F. This could lead to cars and airplanes with plastic bodies. Lear had a plastic airplane—plastic not in our conventional sense of a PVC airplane but very high-strength plastics, epoxies and carbon graphite fibers. More than two hundred buyers lined up pending FAA certification.†

We're sitting on top of a mountain of technological possibilities. New ventures are eagerly trying to make the most of these new opportunities. For now most are small and little known, but some may be tomorrow's winners. How many of these companies do you know? Monoclonal Antibodies Inc., Cytogen, Teknowledge, Cognitive Systems, Robotic Vision Systems, Intermagnetics General, Supercon. Each is exploiting a new technology in a small market niche, just as Texas Instruments did in 1955 when it took on Westinghouse and other established companies.

THE FOURTH WAVE

There is a precedent for what is happening. Historically, technological discontinuities have happened quite frequently but in bunches. The later part of the nineteenth century was, for example, a particularly active time as we shifted from

* Smith, Geoffrey, "The New Alchemists," *Forbes,* April 9, 1984, pp. 101–4.
† But it never came. Lear has now declared bankruptcy.

agricultural to industrial production with the automobile and the train replacing the carriage, the telephone opening up new avenues of communication and steamships replacing the sailing ships. Among economists there is an increasingly accepted belief that waves of innovation have occurred more or less regularly over the past 250 years in roughly 50-year cycles. The first few years see a buildup of new technological potential. These are followed by a period during which new and far-reaching innovations burst on the scene, and then things gradually slow down during a long period of commercialization. Nikolai Kondratiev, a Russian economist, first proposed this idea. It was picked up in the 1930s by the German economist Joseph Schumpeter. Schumpeter showed that the first wave lasted from 1790 to 1840 and was based largely on the new technologies in the textile industry, which exploited the potential of coal and steam power. The second wave took place between 1840 and 1890 and drew directly on the development of railways and the mechanization of production. A third (1890 to 1940) was based on electric power, advances in chemistry and the internal combustion engine. Our current fourth wave (1940 to 1990?) is based on electronics, but the pace of innovation may not pause the way it has between previous cycles. Christopher Freeman, professor of science policy at Sussex University in England, thinks that biotechnology could be at least part of the basis of a fifth Kondratiev wave, which may have already started. I would add advances in physics and mathematics, which underlie advancing computer science, as another key element to the fifth wave.

Change is coming so fast in electronics that my Japanese colleague Ken Ohmae likens it to the fashion industry. Twice a year, if not more often, producers change their offerings. If fashion dictates the electronic equivalent of long skirts and producers come out with short ones, they can be in deep trouble. The Japanese have coined a new word, TAT. It's from the American phrase turn-around time. That is the time between the perception of the need or demand for a new product and its shipment in large quantities. In color TVs, Matsushita holds something of a record—a TAT of 4.7 months. It is a record that probably won't last long.

Another colleague at McKinsey, Steve Walleck, reckons that the use of advanced computer technology (computer-integrated manufacturing) could lead to a 25 percent reduction in the product development cycle for cars. That is an eighteen-month savings. It would allow the automobile companies to make their decisions about which automobiles to produce a year and a half closer to the actual market entry, a tremendous advantage over competitors who fail to use the new technology.

Computer engineers, telecommunicators, machine minders, chemical Ph.D.s, pharmacists, writers, agricultural materials suppliers, fast-food operators, bankers and insurers, arbitrators and lawyers, fashion designers, consultants and even antique dealers are all going to be vulnerable to technological change. With so much change present and ahead, it clearly behooves management to rethink its approach to technology—to have an approach to dealing with Jim Utaski's problem. But for many of them it has been thirty or forty years since they paid real attention to technology.

THE DECLINE OF R&D

Right after World War II the importance and power of technology ranked high thanks to technological advances like radar, encryption, synthetic rubber and, of course, the atomic bomb. U.S. corporations tried to systematically generate and manage new products based on these technologies by building extensive laboratories dedicated to doing "good science." Often these laboratories were located far from the corporate headquarters in order to give the scientists time to think and be free from the "distractions" of senior management. I call this, a bit pejoratively, the "lab in the woods" phase. The popular line of reasoning in corporate boardrooms at the time was that if a company wanted to win it needed to invest in its own variant of a Los Alamos or an Oak Ridge, the labs that had produced our military victory. There they could put their best scientists to work and wait for the results and ensuing profits.

It didn't work out according to plan. The labs received little in the way of specific guidance. There was inadequate interaction between the scientists and their counterparts in marketing and manufacturing. There were few controls on spending, and often no controls on the overall output, perhaps because the lab director often reported directly to the CEO. In truth there was little integration between the general technical efforts and the commercial thrust of the business. Writing of one isolated research center, *Fortune* described it as "holding raucous weekly meetings in a 'bean-bag room' where people tossed around blue-sky concepts while reclining on huge pellet-filled hassocks."* Of course, there were labs in the plants that took care of improving the production process and the handling of product service problems. But the new "lab in the woods" wasn't concerned much with that.

By the late 1950s it was becoming apparent that the promise of easy profits from the technological cornucopia

---

* Uttal, Bro, "The Lab That Ran Away From Xerox," *Fortune*, September 5, 1983.

was not being fulfilled. The R&D community became defensive, isolated and less influential.

Then in 1960 Harvard's Ted Levitt published his classic article "Marketing Myopia," which in effect said, "We've forgotten about the needs of our customers. We need to get back in touch with them." And indeed that was a correct analysis. The age of marketing arrived, bringing with it the second era of technology management. A good name for this phase might be "Marketing Is the Answer." It was based on the notion that if companies understand the needs of their customers they will be better competitors.

Marketing clearly had the upper hand. The role of the R&D vice president was to respond, whenever he was asked, "Yes, I can develop that quickly and cheaply." In terms of planning, performance measurement and control, all were market driven. The period saw our first attempts to deal systematically with the needs of customers, at least the expressed needs. Market research gained respect and importance. Hundreds of different techniques were tried to ferret out what the customer really wanted and feed it to R&D, which was expected to create the new products customers said they wanted. So much was marketing in control that advertising agencies, with no claim to technical knowledge, were frequently put in charge of developing new product concepts.

Inevitably, as the philosophy changed from technology driven to market driven, the relationship between the CEO and R&D became weaker. The technology budget became part of the general budget or even the marketing budget. The R&D officer dropped down the organization chain, and frequently key decisions were taken by the "business team" without him being present. This wasn't by Machiavellian intent. It was often because the business team was physically in the same place, except for the R&D man. He was literally out-of-town in his lab far away from corporate headquarters.

The number of new products accepted by the customers went up dramatically during this period. There were genuine product improvements, but there were also some dubious developments such as toothpaste with stripes and cars with ever larger fins. Unfortunately profits remained flat because everybody was doing the same thing. In their zeal to find out what customers wanted, companies forgot about meeting these needs in unique and competitively protectable ways. Since they couldn't protect new products from competitive imitation they failed to make money.

Somehow this was blamed on technology and it was pushed even farther into the corporate doghouse. By the end of the 1960s, it was not fashionable to be "pro-technology." R&D budgets were slashed. Ph.D.s drove taxi cabs, and companies started to look for new opportunities through diversification. The conglomerate era had arrived.

That is except at a few companies such as IBM, GE, Du Pont, Monsanto, United Technologies and Citibank. Savvier than most, they realized the problem wasn't their technology but the way they guided and managed technology. They saw that to be successful they couldn't make an either-or choice—either technology or marketing. They had to meet the needs of their customers, but in a way that gave them a sustainable competitive advantage. Technology could do that. Rather than downgrading or isolating technology, they had to integrate it into the mainstream of their business so they could use all its potential to beat the competition.

Thus they found themselves in a third era, where technology was seen as a way of gaining and sustaining a competitive advantage. Again the managers of technology development became an integral part of the business team, helping to set, monitor and control their R&D budgets according to the larger needs of the whole corporation. Technology was managed "strategically." As more companies became familiar with this approach they found it worked a great deal better because it focused on meeting the customer needs in new ways that could be protected. But even though this approach to managing technology was clearly better than its predecessors, it still didn't supply all the answers. Leaders still lost. They were replaced by companies with new technologies and more cost-effective products. The strategic approach to managing technology worked for periods of continuous or evolving technical change, but by the time companies started using it we had entered our current period of rapid and discontinuous change.

MANAGEMENT OF DISCONTINUITIES

Companies now need to enter a fourth era. Call it "the management of discontinuities." It will require us to specify and measure technological performance, both output as well as input, and to seek and understand alternative approaches and their limits. It may involve radical shifts in the way many companies are organized and managed that will affect their entire culture and everybody in them. Already a few companies are beginning to implement management systems that seek to do this. According to the *Financial Times*, BMW, the German car maker, is "splitting its research operations from the [new car] development functions. The move was instituted by Hans Hagen, science and research director, who points out: 'There is so much going on in the way of legislation which puts pressure on car development—things such as the emission control rules—that it is better to separate research. With so much pressure on the development functions because of politics as well as normal needs, there is the danger that research might be squeezed.' To ensure [that] essential long-term research will remain available to BMW, a new 'W' department (for Wissenschaft, or science) is being formed. 'But we have no intention of setting up an academic institute—all the work will be product-oriented.' "*

Even the leaders will need to improve how they manage technology to maintain their competitive edge. At the heart of these changes is their need to understand the limits of their present technological approaches.

* "Why BMW Is Untroubled by a Dent in Its Market Share," *Financial Times*, August 12, 1985, p. 6.

THREE

LESSONS
FROM THE
LIMITISTS

I accost an American sailor, and inquire why the ships of his country are built so as to last but for a short time; he answers without hesitation that the art of navigation is every day making such rapid progress that the finest vessel would become almost useless if it lasted beyond a few years.

—Alexis de Tocqueville
Democracy in America, 1840

*I*n 1963, Bob Hayes (later of the Dallas Cowboys) reached a top speed of 28 miles per hour during a 100-yard dash. It is the fastest speed ever recorded for a human. Will anyone run faster? Today nobody really knows because we are not certain what limits the performance of runners. Not that there aren't a lot of ideas. Most center about how much energy can be released from the leg muscles in a short period of time. This is key because the speed at which a sprinter can bring his back leg forward and plant it on the ground is generally felt to be the limiting physical motion. This in turn is limited by the rate at which energy can be released from the muscles. Energy release in the muscles is a chemical process that itself is triggered by an electrical impulse from the brain. The faster the electrical pulse from the brain, the faster the ultimate energy release and the greater the power of the runner. The trouble is we don't understand what limits the brain pulse. So we can't really say how much faster the sprinter can sprint.

Similar analyses have been done for long-distance running where the limiting mechanism is no longer brain trigger rates but the energy in the body's batteries. Unlike those in our cars, human batteries burn glycogen (a long chain of glucose molecules—a kind of sugar). Our batteries are charged up by various chemical processes that start with starches, fats, alcohols and sugars. An understanding of these processes allows sports scientists to speculate how much faster marathoners will run in the future. Some feel that a 2-hour time is possible in the next 25 years—10 minutes faster than the present record.

Understanding limits is as important for businessmen as it is for athletes. In the late 1950s, IBM realized they had a serious problem—a limits problem. If they continued to try to push the approach they were using to design the package of computer chips, new and faster chips would generate so much heat that it would become impractical to keep computers cool enough to run efficiently. Realizing this, IBM developed a wholly new series of computer chips which circumvented the limits imposed by excessive heat. The new chips became the heart of IBM's very successful 360 series.

Subsequently, IBM has used other new approaches to chip air-conditioning as the basis for their design of the 4300 and 308X series of computers introduced in the late 1970s and early 1980s. Both were tremendous successes, and an understanding of the limits of their current technology was a key to IBM's decision to pursue new approaches.

Owens Corning Fiberglass Corporation is not usually thought of as a high-tech company, yet they place a high value on understanding the limits of their technologies as well.

As they told the story to me in 1984, it went like this:

"Several years ago, we began to be concerned that one of our products might be mature since technical process seemed very slow; consideration was being given to cutting back on

R&D funding. Further, two competitive products had picked up market share against ours—one in specialized applications where customers were willing to pay more for a higher-cost product, the other more broadly due to its aesthetic advantages—and we were uncertain as to how far they might go. We decided to undertake an investigation of the technical limits of both our product and the competition's.

"We started by trying to get a precise view of which technical factors of our product were most important to its users. We then sought to determine both the theoretical and practical limits to these factors. For several of the factors, we found we were very knowledgeable and were able, with some effort, to determine the limits. In fact we discovered substantial remaining technical potential and developed several ideas for attaining some of them without serious increases in cost. For one of the factors, we found that having relied in the past on an empirical understanding of the inherent properties of the material, we could not really determine the limiting mechanisms, and we therefore initiated a research program directed toward this end. Research was also undertaken to determine the technical limits of the competitive products. For the broad-based competitive product we found that the technical factors were the same as for our product, but the limits to these factors were different; this gave us a set of research targets for our product. For the specialized competitive product the technical factors themselves were different.

"We found that a systematic identification of the technical performance factors determining the value of our product and a clear view of the mechanisms limiting those factors were very helpful in directing our research efforts. We accelerated our research, to reduce the gap between the performance of our product and its limits and those of the competitive products."* Thus, limits are broadly important to a business whether it be labeled high or low tech.

Limits are important because of what they imply for the future of the business. For example, we know from the S-curve that as the limits are approached it becomes increasingly expensive to carry out further development. This means that a company will have to increase its technical expenditures at a more rapid pace than in the past in order to maintain the same rate of progress of technical advance in the marketplace, or it will have to accept a declining rate of progress. The slower rate of change could make the company more vulnerable to competitive attack or presage price and profit declines. Neither option is very attractive; they both signal a tougher environment ahead as the limits are approached. Being close to the limits means that all the important opportunities to improve the business by improving the technology have been used. If the business is going to continue to grow and prosper in the future, it will have to look to functional skills other than technology—say marketing, manufacturing or purchasing. Said another way, as the limits of a technology are reached, the key factors for success in the business change. The actions and strategies that have been responsible for the successes of the past will no longer suffice for the future. Things will have to change. Discontinuity is on the way. It is the maturing of a technology, that is the approach to a limit, which opens up the possibility of competitors catching up to the recognized market leader. If the competitors better anticipate the future key factors for success, they will move ahead of the market leaders.

* Foster, R. N., et al., "Improving the Return on Research and Development," Industrial Research Institute, December 1984, p. 16.

THE LESSONS OF THE BIG SHIPS

The *Thomas W. Lawson* was a dramatic example of the market impact of being at a technological limit. Clipper ships were clearly approaching their limits 50 to 60 years before the *Lawson* sailed. The fastest of the clipper ships was the *Sovereign of the Seas*. On a trip from London to Sydney in 1854 she achieved a speed of 22 knots (about 24 miles per hour). Other ships in the 1850 period achieved speeds not too different from the *Sovereign's* record. The *James Baines* hit 21 knots in 1856 on a trip from Melbourne to Liverpool. *Defiance* hit 20 knots on a run from Rockland, Maine, to New York in May 1854. It was clear that clipper ships were not getting much faster.

This was the case because of the relationships between the amount of sail flown, the ship's speed and maneuverability. The speed of these ships was governed by the sail they carried, the shape and displacement design of their hulls, and their length. Some observers think that the only difference between the ships of 1850 and those of 100 years earlier was their length. Naval architects had simply made ships faster by making them longer with few other design advances. As H. I. Chapelle said in his history *The Search for Speed Under Sail*, " . . . the gain in speed, in knots, of fast sailing ships of the 1850s over earlier vessels was due to the increase in size, particularly in length, that had developed between 1830 and 1850. The important area of development would then be in construction, allowing wooden ships to be built of great length, with sufficient longitudinal strength to prevent change in form."[*] By about 1850, sailing ships had reached their own natural speed limits, assuming they had to retain

---

[*] Chapelle, H. I., *The Search for Speed Under Sail*, New York: Norton, 1967.

their maneuverability. This was where the *Thomas W. Lawson* failed. Her design attempted to go beyond the limit.

The fact that the limits on sail and speed were known makes an important point about limits. If one understands the underlying science and technology one can estimate where the limits will be. Thus limits, if known and known correctly, confer a degree of predictability on the S-curve that it would otherwise not have. And predictability is what makes the S-curve a useful concept.

If one knows that the technology has little potential left, that it will be expensive to tap, and that another technology has more potential (that is, is further from its limits), then one can infer that it may be only a matter of time before a technological discontinuity erupts with its almost inevitable competitive consequences. Indeed, sailing ships provide us with an example here too. There were alternatives to sail, including steam power, which began to make its debut in about 1840, and even turbine power, which debuted in 1894 with the British *Turbina* (Exhibit 3). It was also known that the speed limits of these boats were far higher than for sailing ships, and that therefore a discontinuity was probably inevitable. It was against this background that the *Thomas W. Lawson* was conceived, built and put into commercial operations. And it was bound to lose.

_____ FINDING THE LIMIT

All this presumes we know the answer to the question "Limits of what?" The "what," as Owens Corning expressed it, was the "technical factors of our product that were most important to the customer." The trick is relating these "technical factors," which are measurable attributes of the prod-

3 The British *Turbina*, 1894.
An experiment with turbine power, the *Turbina* represented one of several well-known new technologies that threatened commercial sailing ships years before naval architects stopped designing new sailing ships.

uct or process, to the factors that customers perceive as important when making their purchase decision. This is often easy enough when selling products to sophisticated industrial users because suppliers and customers alike have come to focus on these variables, for example, the specific fuel consumption of a jet engine or the purity of a chemical. But it is much tougher to understand these relationships in the consumer arena. How does one measure how clean our clothes are? Do we do it the same way at home as the scientists do in the lab? Do we really measure "cleanness," or its "brightness" or a "fresh smell" or "bounce?" All of these are attributes of "clean" clothes which may have nothing whatsoever to do with how much dirt is in the clothes. Or what about diapers? How does the consumer judge those? On capacity? On dryness? On fit? These are complicated questions to answer because different consumers will feel differently about these factors, creating confusion in the lab. Further, once the consumer has expressed his preference it may be difficult to measure that preference in technical terms. For example, what does "fit" mean? What are the limits of "fit"? If the attribute that consumers want cannot be expressed in technical terms, clearly its limit cannot be found.

Further complicating the seemingly simple question of "limits of what?" is the realization that the consumer's passion for more of the attribute may be a function of the levels of the attribute itself. For example, in the detergent battles of the 1950s, P&G and its competitors were all vying to make a product that would produce the "cleanest" clothes. It was soon discovered that in fact the clothes were about as clean as they could ever get. The dirt had been removed, but the clothes often had acquired a gray, dingy look that the consumer associated with dirt. In fact, the gray look was caused by torn and frayed fibers, but the consumer did not appreciate this apparently arcane technical detail. Rather than fight with consumers P&G decided to capitalize on their misperceptions and add "optical brighteners" to the detergent. These are chemicals that reflect light. When they were added to the detergent and were retained on the clothes, they made the clothes appear brighter and therefore cleaner in the consumer's eyes, even though in the true sense they weren't any cleaner.

The consumers loved it, and bought all the Tide they could get in order to get their clothes "clean," that is optically bright. Then P&G reasoned, "If the consumer likes bright clothes, can't we make the clothes even brighter?" And indeed they could. So they did, and more detergent was sold. Once again the question was asked, and once again the technical department succeeded in making clothes optically brighter, but by now a new kind of limit had been crossed. It was not the limit of optical brightness, but rather the limit of the perception of optical brightness and cleanliness. The consumer's desires had been sated. More wasn't better. So even though the technical limits had not been reached from the producers' point of view, they had been from the consumers'.

Another complication with performance parameters is that they keep changing. Frequently this change is due to the consumer's satisfaction with the present levels of product performance; optical brightness in our prior example. This often triggers a change in what customers are looking for. No longer will they be satisfied with optical brightness alone; now they want "bounce" or "fresh smell," and the basis of competition changes. These changes can be due to a change in the social or economic environment as well. For example, new environmental laws (which led to biodegradable detergents), a change in the price of energy, or the emergence of a heretofore unavailable competitive product like the compact audio disc or high-definition TV. These changes in performance factors should trigger the establishment of new sets of tests and standards for the researchers and engineers involved in new product development. But often they don't. They don't because these changes are time-consuming and expensive to make, and they are difficult to think through. Thus it often appears easier to just not make the change. But, of course, this decision carries with it potentially significant competitive risks.

All this assumes that these changes in customers' perceptions of what is important are noticed in the first place. Often they are not. Not at least until a competitor enters with a new product with a new attribute and a new appeal. By then it's often too late for anything other than a competitively weak "me-too" response which has little impact on the market other than a decrease in price. The customers love it. The producers hate it.

As a practical matter, it seems that the people who should be sensing these potential changes in customer tastes, the salesmen, are not particularly well attuned to seeing them. They would rather sell today's products. The people we rely on to keep us close to the customer and new developments often do not. So our structure and systems work to confirm our disposition to keep doing things the same way. As Alan Kantrow, editor at the *Harvard Business Review,* puts it, "Our receptor sites are carrying the same chemical codes that we carry. We are thus likely to see only what we expect and want to see." The chief executive says, "I've done good things. We're scanning our environment." But in fact he is scanning his own mind.

Even if sales and marketing do perceive the need for change, they may not take their discovery back to their technical departments for consideration. If the technical departments do hear about these developments, they may not be able to do much about them because of the press of other projects. So all in all, changes in customer preferences get transmitted slowly, usually only after special studies are done specifically to examine changing customer preferences. All this means that answering the "limits of what" question can be tricky under the best of circumstances, and much tougher in an ongoing business.

_____ *LESSONS FROM THE LEADERS*

If the answer can be found and if companies feel it is important to know the limits, how do they go about finding them? Lew Branscomb, chief scientist at IBM, describes their process:

"At IBM we have an explicit way to deal with this limit. Our corporate research organization identifies all the major technologies on which our business depends—electronic logic technology, for example—and always has two major projects in place for each technology. One is aimed at determining what limits nature sets to the improvement of current mainstream technology. For electronic logic, that would be silicon electronic-circuit technology of ever smaller dimensions. The second project is aimed at the single most promising radical alternative. For electronic logic that would be superconducting Josephenson technology [IBM has cut research on Josephenson junction technology since Branscomb made this comment] or perhaps gallium arsenide technology. Thus we try to have knowledge of where technical limits are —knowledge that is quantitatively expressed and based on actual research—in order to better understand the limits of our business possibilities."*

In principle this sounds like a straightforward exercise, but it can become complex, particularly where the underlying science is new or complicated. It can also involve a good measure of technical and business judgment and intuition. Ian Ross, president of Bell Labs, describes his thinking on the limits of electronic technology:

* "Research & Development: Key Issues for Management," The Conference Board, #842, 1983.

. . .

"The first challenge, of course, is to continue to reduce the size of elemental components. This depends largely upon our ability to improve lithography. Now over the past decades using photolithography with visible light, we've gone from minimum line widths of 25 microns—1/1000 inch—down to the present industry average of 2.5 microns—1/10,000 inch—approaching the wave length of light. But we're shortly going to run into the limit of the wave length of visible light and that will be a problem. It is estimated that, using visible light and all the tricks we can conceive of, we might ideally get down to line widths of a half micron. Under practical conditions, however, this is more likely to be 1 micron.

"The ultimate technique though, one that has already been used, is to use electron-beam lithography. And there one can predict, and indeed can demonstrate, that lines within the range one-tenth to one-hundredth micron are possible.

"Now when we get into that range, we are dealing with distances less than the spacing of a hundred atoms, down to ten atoms. Here, one wonders if there are not some other limitations than one's ability merely to produce small lines. And, indeed, there are.

"That more restrictive challenge is making operating devices that small. And here we find that there is a fundamental limitation in silicon as in other materials, and that is related to dielectric breakdown strength. Just as air breaks down under high electric field to create lightning, so too does silicon. The ultimate dielectric breakdown strength in silicon is about 100 thousand volts per centimeter. To make a useful silicon device operate at room temperature, therefore, you need to apply to it electric potentials in the neighborhood of one volt. And leaving a reasonable factor of safety, that translates into structures with dimensions no less than about a tenth of a micron. This then is more limiting than lithographic capabilities.

"In the lab today, transistors with critical dimensions of 0.1 micron have been made, have been shown to operate, and have been shown to perform according to theory. So possibly we may be able to achieve structures of about a tenth micron minimum dimensions with about a hundred atoms within those minimum dimensions. This could lead to a mere billion components on a square centimeter of silicon. That is not very restrictive."

But Ross goes on to explain that there are other, even more severe restrictions on the number of components we can place on a chip (which affects the performance on the chip). They include the present approach of light-based circuit printing, but a more fundamental limit is imposed by the molecular structure of chip material—silicon. Ross then points out an even more restrictive limit—the old problem of connection.

"My argument so far considers only the minimum size of an operating component itself, about three times the minimum dimension that can be contemplated. Not included are the requirements to interconnect these components with current-carrying conductors that will bring adequate power to the devices. Nor does it include the minimum separation of elemental components necessary to provide adequate isolation. When these factors are taken into account, a more realistic upper limit may well be 100 million components per square centimeter.

"Having dealt with minimum size of components and hence density on a chip, what can be said about future increases in the size of the chip itself? How many square centimeters per chip? This ultimately is limited by manufacturing control—the ability to minimize the size and density of defects. The leading-edge chip-size today is about one square centimeter. There is speculation about chips the size of a silicon wafer, so-called 'wafer-scale integration.' This would give a factor of 100 increase in chip size. Now frankly, I am neither bold enough to say that this can happen in the foreseeable future, nor am I rash enough to suggest that no progress will be made in this direction.

"So for today, with considerable discomfort, I will settle on an estimate of 10 square centimeters—about an inch square—maximum obtainable chip size. This area, together with a density of 100 million components per square centimeter, would give the ultimate goal of one billion components on a chip of silicon. A startling result but difficult to ignore. [Author's note: This would be like having 1,000 IBM PC's on a single chip a little more than 1 inch on a side. We could have in our homes all the computer power of a major space installation or a bank. We will probably use such devices for video or speech synthesis.]

"Now if we accept the complexity limit, the next question we have to ask is how far can we go in increasing the speed of these devices. Because, clearly, as the components get smaller, they will operate faster. Here the ultimate limit is bounded by the maximum velocity at which electrons can move in silicon. And if you know that, then of course time is equal to the distance divided by velocity. The maximum velocity in silicon under normal conditions is about 10 million centimeters per second, or one-thousandth the speed of light. So given the critical minimum dimension of a tenth of a micron we have talked about, you can simply calculate an ultimate switching device speed of about a trillionth of a second—a picosecond (10^{-11}). [A picosecond is a very short period. To put it in perspective assume that the universe is about 10 billion years old. A picosecond is to a second, as 2½ days are to the life of the universe! Not very long at all.]

"Here again there are some circuit limitations to consider, in this case the capacitive loading of not only the components themselves, but importantly of their interconnections. When you take these factors into consideration, a more realistic estimate of ultimate speed is 10 picoseconds -10^{-11} seconds (or about one month in the life of the universe).

"So having gone through all those physical principles, and all that arithmetic, we come to this startling conclusion: there are no fundamental barriers to making silicon integrated circuits on chips about an inch square and containing as many as a billion components, each operating in a time as short as 10 picoseconds."*

It now seems certain that Bell Labs has revised their notion of Ian Ross's speed limit. As Ross mentioned, the switching speed is limited by the maximum speed of an electron in silicon. It turns out that this speed is limited by the number of silicon atoms the electrons might bounce into on its trip. If that number approaches zero the electron can zip through the silicon at a much faster speed than Ross's 10 million centimeters per second. The electron essentially flies through the silicon the way a bullet flies through the air unrestrained until it hits its target. In this case the electrons are said to travel by "ballistic transport," and the devices made to exploit this phenomena are called "ballistic transistors." Bell Labs has done a computer simulation of the ballistic transistor† and finds that it would have switching speeds of 10

---

* Ross, Ian, "Limits of Semiconductor Technology," Sixth Mountbatten Lecture, London, November 1983.
† *Business Week*, February 11, 1985, p. 32.

femtoseconds or 100 times faster than the fastest transistor Ian Ross described above! The Japanese have recognized the significance of the development. It is reported* that the Japanese electronics giant Fujitsu will be producing a commercial ballistic chip in the spring of 1986 for low-noise amplifiers for satellite communications. It will allow a 6.5-foot-diameter disk to be shrunk to 2.5 feet. Fujitsu's chip will switch 20 billion times per second, but even this incredible rate is far from the limits for ballistic transistors.†

LIMITRY

What Ross demonstrates here is a way of thinking. He talks about fundamental principles that will ultimately stop progress. I call them limiting mechanisms. They are the qualitative descriptions of what will stop progress, like the air breakdown phenomenon that results in lightning. Identifying these limiting mechanisms rests with scientists and engineers who understand the underlying science, if indeed it is understood. If it is not, then it needs to be discovered through focused laboratory work. The objective of the research is to find limits, whether they tell us bad news or good. Success is defined by the act of finding the limits. Failure is the failure to find

* Ibid, September 30, 1985, p. 87.
† As a postscript to Ian Ross's comments, it is interesting to note that Gene Amdahl, ex-IBM chief engineer on the IBM 360 and 370 computers and founder of Amdahl, founded another company called Trilogy to design and develop "wafer scale" integrated circuits. These are the ones that Ian Ross referred to as 100 times as big as the present chips. If wafer scale integration was accomplished, one could ultimately put 10 billion components on a chip. But Trilogy hit snags in the development. They could not find a way to dissipate the heat generated by the chips. So investors who bought in at $12 per share, now hold stock worth about $1 per share. Maybe someday we will see Gene Amdahl's wafers, but they probably won't be from Trilogy.

the limits. Failure is not the failure to find the kind of limit we would like to have. A researcher makes progress both by trying to think through to the end of the problem, that is the ultimate limit, and by working his way forward from where the technology is today, conceptually solving one problem after another, or judging whether it can be solved, until he hits one he knows can't be solved, because he knows the fundamental science. So he goes from the science backward or from the technology forward, using judgment along the way until he understands the network of limits that precede any technology development.

Ian Ross is particularly good at this art. He is a brilliant and experienced scientist, has a good understanding of what the business problems are, and understands the constraints placed on any product by the way it is produced. He has become particularly skilled at seeing the limits of technologies where others do not—a limitist as we have defined the term.

Try a test case yourself. Look at the eighteenth-century rotating device in Exhibit 4. This clever device was supposed to continue to spin endlessly once it got started. A heavy lead ball at point A rolls down the incline plane B until it lands in cup C. At that point it exerts such a force on arm D that it causes the axle in the center of the wheel to turn. As this happens the balls on the other side of the wheel are raised, until one reaches point A and the process starts over again.

But this machine doesn't work. Why not? Look at the picture again before reading on.

There are two answers. First the wheel is required to lift the ball to point A which is higher than point C. This is like water running up hill. It can't be done without an external source of energy. Second, the wheel as pictured requires the two balls on the right side of the wheel to pull up five balls

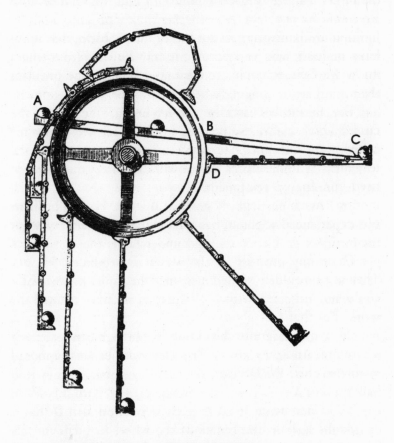

4 An Eighteenth-Century Perpetual Motion Device.
It was designed by George Lipton to spin endlessly. It didn't. It couldn't.
Source: Ord-Hume, Arthur, W. J. G., *Perpetual Motion*, New York: St. Martin's Press, 1980.

on the left side, and this can't happen since the two balls
weigh less than the five balls. If the machine ever got started
it would probably turn in the opposite direction from the one
intended by the inventor.

Even if you had a technical education you probably

didn't get the right answer at first or it took you a while. Ian Ross saw the problem immediately. He is used to looking for limiting mechanisms; most of us are not. Most corporations have limitists who can be of critical value. But they don't know who these people are because they haven't looked for them.

The absence of limitists can be costly. The wheel discussed above is an example of a whole class of machines called perpetual motion machines. They purport to be able to generate more energy than they consume, removing the need for energy sources altogether. If we had perpetual motion machines we wouldn't need oil, coal, water power or nuclear plants, or the corporations that provide them.

These machines do not exist because they can't exist. They are forbidden by the laws of physics. Just as two balls could not lift five, neither can perpetual motion machines work. But that fact doesn't mean that there aren't plenty of people trying to ignore the existence of this particular limit. There are. And they are all very interested in raising money to further their R&D.

(UPI) December 14, 1984: A Mississippi man who claims he has invented an energy-producing machine received a promise of more than $10 million from three Californians who say the device will mean "a new day in history." Joseph Newman of Lucedale, Miss., said he has spent 19 years developing the prototype of a machine he believes eventually could make individual homes and businesses energy self-sufficient at a minimal cost. The device uses an electromagnetic force to continually produce energy, Newman said, and the prototype emits at least 10 times as much energy as is put in. Eventually, he said, the output of energy could be infinite.

Needless to say this machine would have a ready market, even at a heady price, were it not for the fact that Mr. Newman's machine very likely violates the laws of physics. Ignorance of the fact will not change the fact. And this may be unfortunate for the three Californians reported to be putting up the "more than $10 million," one of whom was reported to say, "There will be no force on this earth that is going to stop it. Just imagine . . . an environment free of nuclear reactors, an environment free of oil, coal and gas. It will be a new day in history." Such investor savvy will not change the laws of physics, nor enable machines to operate beyond the limits.

Another UPI story, dated February 7, 1981, reported that Arnold Burke was found not guilty of perjury in a Texas court. Burke had invented an electricity-generating machine which he called "Jeremiah 33:3," referring to the Biblical text, "Call unto me and I will answer thee, and show thee great and mighty things, which thou knowest not." The perjury charge stemmed from a deceptive trade practices suit brought against Burke by the Texas Attorney General's office where Burke had claimed under oath that the machine would produce power without an outside energy source. During the course of the trial investigators from the Attorney General's office testified that they discovered a hidden wire that led from the heart of Jeremiah "to a battery pack in another room and eventually to an electrical outlet." Burke's defense was that he had removed the special pumps and installed electrical pumps for the examination to protect his yet-to-be-patented machine. The jury found reasonable doubt of perjury.

The heart of Burke's machine was a submergible energy-free pump that Burke claimed continuously circulated water over an electricity-generating turbine. Water flowed from a 200-gallon tank, through a series of pipes, over the turbine and into a collector near the bottom of the 12-foot-tall machine where special pumps, powered by the turbine, pushed the water back to the tank at the top to start the process again. Excess power from the turbine could be used to produce 3,000 kilowatt hours per month, more than enough to heat the average size home free, if it worked.

These attempts to circumvent the laws of physics are always interesting and sometimes amusing to those of us who are not investing in them. But the seekers of perpetual motion machines are trying to solve a serious problem, even if their approach ought not to be considered seriously. They are really trying to find ways around limits, and this is a valuable undertaking. People who can figure out how to avoid limits, once we know what they are, are the "limit breakers" and are very valuable people.

LIMIT BREAKERS

Jack Kilby solved a limits problem and in doing so invented one of the most important products of the twentieth century —the integrated circuit. The setting is late 1958, seven years after the dawn of the age of transistors. Transistors were the microminiature equivalents of vacuum tubes. They were the little devices that provided an opportunity to make it big to companies like then fledgling Texas Instruments. These miniature electronic components whet everyone's appetite for what could come in the future—the ultramicrominiature components that would open the way for the broad-scale consumer use of electronics.

But there was a problem of limits. To be useful these little components had to be wired together. The more complex the function of the device, the more of these components the engineer had to wire together. But how? A low-fi radio could be quite simple, involving only a few components. But a radio telephone would be very complex and require a lot of components. Jack Morton, vice president for electronic components at Bell Labs, put the problem this way: "[Each component] must be made, tested, packed, shipped, unpacked, retested and interconnected one at a time to produce a whole system. Each element and its connections must operate reliably if the system is to function as a whole." This sets up a "tyranny of large systems, a numbers barrier (limit) to future advances, if we must rely on individual discrete components for producing larger systems."*

Many companies including Texas Instruments were trying to figure out a way around this barrier. TI had thought that if they could make each component a standard size it might help. In fact, it would probably help a little, but the approach still ran into the same limits as had prior approaches. The more complex the device, the more connections that had to be made. What was needed was an "AHA!"

Martin Gardner, a longtime writer of brain teasers for the *Scientific American,* coined the phrase, the "AHA!" experience. It is what you say when you see a problem in a wholly new way. A way that gives you wholly new insights. It's when you see the hole and not the donut. As the famous biologist Albert Szent-Györgyi once remarked, "Discovery

* T. R. Reid, *The Chip: The Microelectronics Revolution and the Men Who Made It,* New York: Simon and Schuster, 1985.

consists of seeing what everybody has seen, and thinking what nobody has thought." That's the "AHA!" experience.

Jack Kilby, as a young engineer at TI, had a different approach to the connection problem. He observed that silicon, the material that had come to be used for the transistors, could be used for the other components as well. They would be more expensive, but, Kilby reasoned, "So what if it costs me more to make these other components out of silicon if by making all of them that way I can put them together without wires?" Without wires, indeed! (AHA!) No more individual components. No more testing and retesting. No more packaging and unpackaging. No more "tyranny of numbers," because all components, no matter how complex, now became one. The problem went away. Jack Kilby had just invented the "integrated" circuit.

Kilby's story is more dramatic than most, but the experience is not atypical for scientists who succeed in circumventing limits. They become obsessed with the problem. They seek radical alternatives but ones that are essentially simple, elegant (make everything out of one material) and potentially low-cost. They use experience drawn from other areas, rather than continue to try to solve the problem with reapplication of the same principles (make the wires smaller, cut the number of connections). As a result, limit breakers often come from fields that are different from those where prior experience has been concentrated. Indeed, that is the advantage of limit breakers. They are not trapped by the decreasingly effective approaches of the past. With their success and new skills comes the need for more new skills in order to exploit the advance. Not constrained by the need to utilize increasingly marginal resources, and free to use the best available new resources, the innovator—in business, the attacker—

gains the potential for advantage at the very point in time when the limit breaker has his AHA! experience.

It is important to recognize the limits of limit analyses. There are two major ones I can think of. The first is that it does not follow that if you are close to the limit of a technology, there is no more technical room to maneuver. What does follow is that there is no more room to maneuver in that technology. That is not to say there are not, or won't be, alternative technical approaches to meeting the customers' needs that have plenty of room for improvement. In most cases there will be. Recall what P&G did for the diaper business when they came out with Pampers. That was a wholly new technical approach to an age-old problem. Now it's rumored that they are planning to take another step forward with a super absorbent material that will allow the diaper to hold much more liquid than in the past. So there are always new ways to do just about anything.

The point is, it doesn't follow that just because you've reached the limit of one technology there isn't another technology that can solve the customer's problem in a superior way. If there is an alternative, and it is economic, then the way the competitors do battle in the industry will change.

There is a second limit to limits, and that is that it's possible to be wrong about what the limits are. If one is wrong about the limits, he will certainly draw the wrong business conclusions and decide that something is possible when it isn't or vice versa. There are, of course, famous and now perhaps funny stories about these errors in the past. For example, about 1900 Simon Newcomb, the celebrated *fin de siècle* astronomer, said, "The demonstration that no possible combination of known substances, known forms of machinery and known forms of force, can be united in a practical machine by which men shall fly long distances through the air, seems to the writer as complete as it is possible for the demonstration to be." Two years later in 1902, Newcomb added, "Flight by machines heavier than air is unpractical and insignificant, if not utterly impossible."

The next year the Wright brothers succeeded at Kitty Hawk and in doing so disproved Dr. Newcomb's proof.

Another example is Dr. Robert Millikan, a Nobel Prize winner in physics, who said in 1923, "There is no likelihood man can ever tap the power of the atom." There are hundreds of similar examples. These men, men of science (but not necessarily knowledgeable in fields outside their own), had used the arguments of limits to convince themselves and others of exactly the wrong conclusions. They had misapplied their science. They lacked the vision, or the knowledge to see around the next constraint. They would have probably used the "tyranny of numbers" to reach the wrong conclusion about the potential for miniaturizing the transistor.

Calculating the limits is an important but potentially difficult exercise. Doing it will require knowing what is not known as much as what is known (did Newcomb "know" that flight was *prohibited* by the laws of nature? Did Millikan know that there were *no* elements that were capable of giving up their energy?) It requires facts, based on experiment. It requires rigorous logic to rule out other possibilities. In short, finding the limits is a very challenging intellectual exercise, but as the IBM case demonstrates, when it is well done, it is an extremely powerful competitive weapon.

Peter Brancazio says about the high jump: "Many scientists believe that there is a definite upper limit to the height attainable in the high jump, based on such considerations as maximum muscle forces, bone stresses and power generation. Today's high jumpers may be very close to this limit.

Between 1952 and 1962, the record for the high jump advanced from 6 feet 8 inches to 7 feet 5 inches, but in the past twenty years it has increased only a few inches more to 7 feet 9 inches. Will anyone ever jump 8 feet? I would be willing to bet it won't happen in this century."*

Brancazio is probably right unless there is a limit-breaking high jumper out there with a totally new approach. And that is the most important question a company must ask its scientists to answer, whether or not there is a limit breaker on the horizon.

* Brancazio, P. J., *Sportscience*, New York: Simon and Schuster, 1984.

FOUR

THE S-CURVE: A NEW FORECASTING TOOL

It is quite impossible that the noble organs of human speech could be replaced by ignoble, senseless metal.

—Jean Bouillaud, member of the French Academy of Science, regarding Thomas Edison's phonograph

*I*n December of 1982, at the University of Utah Medical Center in Salt Lake City, Dr. William DeVries replaced Dr. Barney Clark's diseased heart with an artificial one. Not a metal heart to be sure, but an artificial one that performed nobly none the less. In one sense it was the culmination of work begun many years earlier. But in another sense it is the story of the bottom end, or ascending portion, of the S-curve. It provides insight into how an S-curve looks and feels to those who are trying to push it ahead.

The story begins in 1957 in the Cleveland Clinic. The Cleveland Clinic is a referral hospital for special problems, especially those of the heart. It is a sprawling complex of a dozen buildings spread over 100 acres on the edges of downtown Cleveland. Patients have included President Sekou Touré of Guinea, King Hussein of Jordan, the King of Bhutan, the royal family of Nepal and the King of Saudi Arabia.

It was at the Cleveland Clinic that Willem Kolff and Tetsuzo Akutsu began their research with artificial hearts. In one of their first experiments, they used a plastic sack for a heart to keep a dog alive for 90 minutes (Exhibit 5).

They then began using larger animals, but the first artificial hearts they developed had two major problems. First, they were externally squeezed to pump the blood; the squeez-

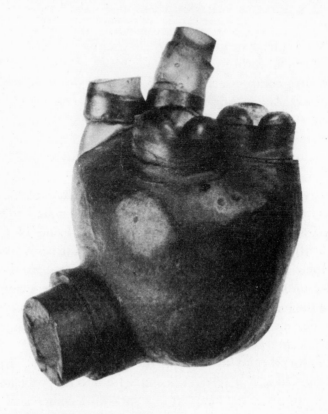

5   Kolff and Akutsu Heart, 1957.

ing crushed most of the blood cells (which doctors call hemolysis) that passed through them. Those cells that survived
saw the heart as an invader in the body and tried to defend
the body by bringing white blood cells to the heart. The
resulting coagulation stopped the blood flow. The animals
all died quickly after the artificial hearts were implanted.

The doctors tried new design concepts to overcome these
problems, but their attempts did not find quick success. As
Dr. Robert Jarvik recalled in 1982:

Over the next few years Kolff's group at the Cleveland Clinic developed and implanted several other kinds of artificial heart driven by electricity. One such device employed five solenoids that displaced oil; the oil in turn compressed polyurethane sacks that held the blood. Animal survivals of three hours were obtained with this heart. In another electrically driven heart an electric motor drove a roller that compressed a blood-carrying tube against a foam-lined housing. The heart needed only outflow valves, but it caused excessive hemolysis and sustained life for only two hours. In the pendulum heart a pivoting electric motor alternately compressed two blood-containing sacks, thereby forcing blood out of the ventricles. Several dogs survived for from four to six hours with this device, but its output was inadequate and it caused excessive hemolysis.*

Other approaches were also tried. One used a nuclear-powered impeller that pumped the blood (Exhibit 6), but the impeller also crushed the cells. Another attempt had rubber sacs activated by compressed air (Exhibit 7). It too crushed the cells. Then, in 1970, Clifford Kwan-Gett, who had been with Kolff at the Cleveland Clinic and moved with him to the University of Utah in 1967, solved the problem by gently pumping the blood with a diaphragm (Exhibit 8).

But the uncrushed blood cells still coagulated on the surface of the heart. While better than the Kolff-Akutsu heart, the Kwan-Gett wasn't good enough.

Another researcher, Dr. Nosé, solved this problem by

* Jarvik, R. K., "The Total Artificial Heart," *Scientific American*, January 1981, p. 77.

6 Artificial Heart.
Developed by Westinghouse and the University of Utah, 1975.

7   Yukihikio Nosé Heart, 1965.

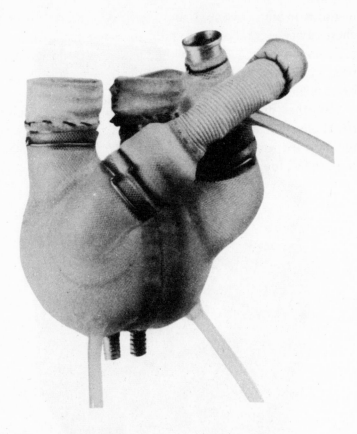

8   Kwan-Gett Heart, 1967.

replacing Dr. Kwan-Gett's synthetic silicon heart interior with natural tissues (such as the outer heart membrane). This approach solved the clotting problem, but blood leaked from this heart because it did not fit well into the chest cavity. As happens so often in the early stages of research, one removes one impediment only to find another. And that is why the S-curve is so flat in the beginning.

Dr. Jarvik, who as a high school student was improving

methods to staple wounds shut, designed a heart that fit the chest cavity much better (Exhibit 9). The experimental animals now lived much longer than before. Real progress had been made. Jarvik began to ascend the S-curve of artificial heart technology.

Another problem struck. Now that the heart kept patients alive longer, the diaphragms began to give out. Jarvik replaced the rubber diaphragms with Lycra, an elastic material used in bras and girdles. It had the necessary wear resis-

9 The Jarvik-7 Heart.

tance. The heart life increased up to four months and later to six months. Subsequently, Jarvik replaced the Lycra diaphragm with a specially designed rubber, and that lasted even longer.

With this evidence, Dr. Kolff and the University of Utah submitted an application for approval to implant an artificial heart in a human in February 1981. After an initial rejection by the FDA in March (for reasons that had nothing to do with the heart's performance but rather with questions about the psychological impact of the implanted heart on the patient), the application was approved in June. Then in 1981, Barney Clark had his heart replaced by DeVries. He lived for 112 days before he died of pneumonia. William Schroeder received the next artificial heart on November 25, 1984. He is surviving but is depressed, as the FDA feared he might be. But he is alive.

With each trial the performance of the heart became better. When the heart's performance, measured by the post-operation lifetime of the patient, is plotted versus the effort the various medical teams put into improving the hearts, the beginnings of an S-curve appear (Exhibit 10). With each subsequent heart representing a new data point, progress is slow at first but then accelerates rapidly as Jarvik makes his advances.

For each new product (or process) the S-curve shows precisely how much performance has improved and how much effort has been expended to gain that improvement. What the development of artificial hearts shows is that S-curves, though they are abstractions, can also be graphic histories of human efforts to solve problems. They represent the trials and errors of talented people. Displaying their efforts allows us to see patterns of success and failure that we

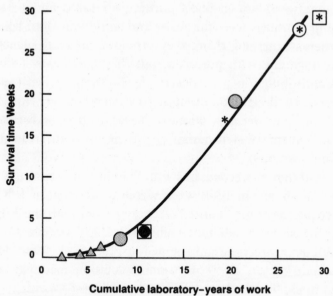

10   Artificial Heart Research.
The early stage of the S-curve for artificial hearts—it represents the inevitable trials and errors of research and invention.

wouldn't otherwise find. In this case a rather long period of little progress followed by growing success.

_____ DIMINISHING RETURNS

The process can also work the other way as technological limits are approached. Rather than showing more and more progress with less and less effort, each new step makes less and less progress. One of the most interesting cases to me of diminishing returns took place over a century ago and involved pocket watches. The "pocket" watch, which was itself something of a minor revolution (clocks were meant to be in towers, not in one's pocket), appeared about 1600. The first models were the shape and size of a lemon. For the gentleman carrying them, they provided the convenience of knowing time with precision, but they did create a rather unsightly bulge in his trousers. Accordingly, it became the fashion to design, or attempt to design, thinner watches. Put in our terms, thinness became the performance measurement. Watch designers put in effort to create ever thinner watches.

And they succeeded, but with diminishing returns, as Exhibit 11 shows. In 1700 when watch production, which is a surrogate term for "effort," was quite low, both the French and British were making watches about $1\frac{1}{2}$ inches thick. By 1800, as production had stepped up, they had gotten down to about $\frac{3}{4}$ inch. By 1850, watches could be made about $\frac{1}{4}$ inch thick. That's the thickness of my "wrist" watch today! Not much different from 1850. If we take the nineteenth-century continental gentleman's definition of performance, thinness, its limit was reached in 1850 and before that came diminishing returns. Each new design represented efforts to pack more and more components into smaller and smaller spaces. As thinness reached its limits, other performance parameters (area, reliability, easy use, and cost) gained in importance.

These two stories, artificial hearts and pocket watches, when put together describe the two major phenomena behind the sinewy shape of the S-curve: learning (artificial hearts) and diminishing returns (pocket watches). The S-curve traces out the path of development of new products and processes with each successive point on the curve representing an improvement in performance. The pattern of the S-curve repeats itself again and again in industry after industry. These empirical observations, coupled with the underly-

| c. 1700 | 1812 | c. 1815 to 1825 | 1846 | 1850 |

11   British and Swiss Watches.
Pocket watches started reaching the limit to their thinness around 1850.
Other performance parameters such as reliability, easy use and cost then
gained in importance.
Source: Landes, David S., *Revolution in Time,* Cambridge: The Belknap Press of Harvard
University Press, 1983.

ing theory of why it is happening, seem to me to be
convincing evidence that these curves describe reality and
will continue to do so in the future.

A FORECASTING TOOL

If that is true and if the limit for an S-curve can be predicted,
then the S-curve can yield valuable insights. If we can define
important performance parameters, trace the early days of
progress of these parameters versus the effort to make the
progress, and develop a point of view on what the limits of
these performance parameters are, then we have a basis for
foreseeing how much further current products can be im-

proved and how much effort it will take to get them to higher levels of performance (see Appendix 2). If S-curves of potential competitors are also sketched, then we can gain some insight about them. Equally, it will give insights about how products will fare in the future, what new products to try to develop, and how much effort will be required to develop them.

S-curves have been constructed for electric power technology, the accuracy of clocks, increasing the efficiency of electric light bulbs, ammonia production, drug dosage, telecommunications band widths, organic insecticides and software and dozens of other technological developments. I've even seen one for "travel," which includes walking, wagons, railroads, cars and airplanes and covers a long period of time. But in all these cases what we want to know is the relationship between effort put in and results achieved. You might think you should be plotting results against the amount of time involved. But that would be an error. It is not the passage of time that leads to progress, but the application of effort. If we plotted our results versus time, we could not by extrapolation draw any conclusion about the future because we would have buried in our time chart implicit assumptions about the rate of effort applied. If we were to change this rate, it would increase or decrease the time it would take for performance to improve. People frequently make the error of trying to plot technological progress versus time and then find the predictions don't come to pass. Most of the reason for this is not the difficulty of predicting how the technology will evolve, since we have found the S-curve to be rather stable, but rather predicting the rate at which competitors will spend money to develop the technology. The forecasting error is a result of bad competitive analysis, not bad technology analysis.

Thus, it might appear that a technology still has great potential but in fact what is fueling its advance is rapidly increasing amounts of investment. Gordon Moore, President of Intel, says that the density of circuits on a chip will double every two years. That has been true, but how long will it continue to be true? If we plot progress against time, it looks as if progress is getting more rapid. But if we plot chip density as a function of effort, there are signs that the technology is beginning to approach its limit. The rate of effort (dollars per year) has been increasing even more quickly than chip density. This means it is getting more and more expensive to develop each new generation of electronic memories. It took about $100 million to develop the future workhorse of the personal computer—the 256K RAM. What will it take to develop the next generation RAM—the million bit (or megabit) RAM? Probably a good deal more than $100 million.

STRATEGIC MANAGEMENT OF DISCONTINUITIES

Ascending an S-curve is something like driving up a hill. You often see caution signs saying 10 percent or 30 percent grade, depending on the hill's steepness. We can think of the slope of the graph just the way we can think about the slope of a hill. The higher the slope, the more productive we are. Thus, a convenient way to pinpoint where we are on our curve of results versus efforts is to talk about the slope or productivity of the technical effort.

At the start of the curve we need to put in significant effort before we can expect to see results. Once the learning is done, we begin to make significant progress for very little expenditure of effort. That usually does not last too long— perhaps a few years. At some point we begin to approach the limits of the technology and we start to run out of steam. Then the question is might there be another way to deliver the desired performance to customers. Some new technology which, though still undeveloped, might eventually outperform the current one, which is increasingly resisting improvement?

But too often we don't ask those questions. Behind conventional management wisdom is the implicit assumption that the more effort put in, the more progress that results. In fact, this is only the case in the first half of the S-curve. In the other half it is wrong. To compound things, it is hard to see what is happening when it is happening because most companies do not keep records of technological productivity.

S-curves almost always come in pairs (Exhibit 12). The gap between the pair of S-curves represents a discontinuity—

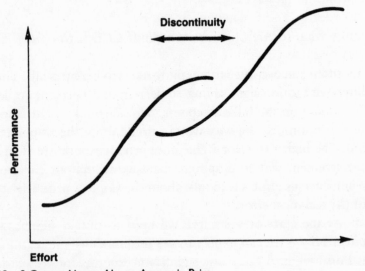

12 S-Curves Almost Always Appear in Pairs.
Together they represent a discontinuity—when one technology replaces another.

a point when one technology replaces another. For example, when solid-state electronics replaced vacuum tubes.

In fact, rarely does a single technology meet all customers' needs. There are almost always competing technologies, each with its own S-curve. Thus, in reality, there may be three or four or more technologies involved in a battle, some on defense and some on offense. Often several new technologies vie with each other to replace an old technology in a market segment—for example, the way compact disc players are competing with advanced tape decks and super-refined turntables for a share of the home stereo market. Deciphering a discontinuity's S-curves when all this is happening is very difficult. Not surprisingly, a friend of mine in the pharmaceutical industry says these periods of discontinuity are "chaos." Indeed, they are.

Companies that have learned how to cross technological discontinuities have escaped this trap. They invest in research in order to know where they are on relevant S-curves and know what to expect from the beginning, the middle and the end of these curves. A few draw very precise S-curves, but it's often enough just to know the general dimensions and limits and accept the implications.

This is what I referred to in the opening chapter as the fourth era in the management of technology—the management of discontinuities. Most companies are in the third era, the one I labeled as "strategic" management of technology. They have become very sophisticated at massaging the shape of the curve, making it steeper by developing new products and processes faster than their competitors. For example, my partner Ed Krubasik in McKinsey's Munich office has studied a number of cases involving faster than normal development for individual products. His hypothesis is that the need to cut development times depends on both the costs of development and the profits that might be missed if development were delayed. Often one or both of these can be substantial. He studied several cases to illustrate his point: The IBM PC, the Boeing 767, the Canon PC 10, and Northern Telecom's "digital switch" (solid-state digital telephone exchange). All had either high technology development costs (Boeing), or high market opportunity costs (IBM PC, Canon PC-10), or both (Northern Telecom). What he found was that the cost of being late to the market overwhelmed the cost increases for accelerating development. In each case, special tactics were used to speed up development.

Design and engineering were done as far upstream as possible. For example, Boeing identified, designed and tested things such as composite tail and wing materials before it went to prototype, even to the point of doing wind tunnel tests. In the cockpit Boeing tested several key components as back-up systems in its 737s before going to final design.

Some companies have also learned how to share the development of technology and products. For its personal computer, IBM bought its monitor from Matsushita, its floppy disk from Tandon, its microprocessor from Intel, its printer from Epson, and its operating system from Microsoft. All companies make extensive use of external suppliers, but not too many manage them efficiently. Yet this is a fertile area for saving time and raising quality control. The same is true for customers. The better companies collaborate as much as they can, and often put specialists on site at their customers' plants.

Administrative procedures can also be eliminated to speed up development. Most new IBM products go through a rigorous eight-phase process. But the IBM PC bypassed this, reporting directly to Chairman John Opel. Companies are learning how to use computers and improved communications to speed up development.

Perhaps most important, companies have learned that in order to be fast to the market, they must invest in understanding the science that supports the base of the S-curve. Too many companies develop products empirically. They know things work, but not why they work. They rush through the engineering and then hit some major problem that requires an understanding of the supporting science which they don't have. There is no base for understanding limits, anticipating progress or fixing performance problems as they inevitably occur when a product is developed too quickly. A thousand engineers get held up while frantic calls go out to basic research.

The RAND Corporation, a government-sponsored think tank headquartered in Santa Monica, California, looked into the cost overruns of "Pioneer Process Plants" in 1981.* They noted:

1. Severe underestimation of capital costs is the norm for all advanced technologies.
2. The factors that account for poor cost estimates and poor performance can largely be identified early in the development of the technology long before major expenditures have been made for detailed engineering, much less construction.
3. Seventy-five percent of the cost variance can be attributed to insufficient technical information *before* the project began.

Said more simply, companies that get products to market ahead of their competitors still don't take shortcuts: they do

---

* Merrow, E.W., et al., "Understanding Cost Growth and Performance Shortfalls in Pioneer Process Plants," RAND, 1981.

their research first, then tackle engineering. As a result their S-curves are steeper than other companies trying to develop the same technology.

But it is important to understand that all of these things, while they can be used to improve the productivity of R&D expenditures and get new products to market fast, can be futile. As futile as NCR trying to sell its new electro-mechanical cash register to customers when they were awaiting electronic ones. None of these efforts will save a company from a new technology.

EFFICIENCY VERSUS EFFECTIVENESS

Superb "third era" management of technology has one major problem: it focuses on efficiency when companies need to be concerned with effectiveness. Effectiveness is set when a company determines *which* S-curve it will pursue (e.g., vacuum tubes or solid state). Efficiency is the slope of the present curve. Effectiveness deals with sustaining a strategy—efficiency with the present utilization of resources. Moving into a new technology almost always appears to be less efficient than staying with the present technology because of the need to bring the new technology up to speed. The cost of progress of an established technology is compared with that of one in its infancy, even though it may eventually cost much less to bring the new technology up to the state of the art than it did to bring the present one there. To paraphrase a comment I've heard many times at budget meetings: "In any case the new technology development cost is above and beyond what we're already paying. Since it doesn't get us any further than we presently are, it cannot make sense." The problem with that argument is that someday it will be ten or twenty or thirty times more efficient to invest in the new technology, and it will outperform the existing technology by a wide margin.

There are many decisions that put effectiveness and efficiency at odds with each other, particularly those involving resource allocation. This is one of the toughest areas to come to grips with because it means withdrawing resources from the maturing business. What makes this decision so difficult is that it is being made from the inside and resources will have to be withdrawn from businesses that in many cases sired the CEO, or from the division that has been the major building block for the company and has the strongest political ties to its current management.

In addition, many companies have management policies that, interpreted literally, impede moving from one S-curve to another. For example, "Our first priority will be to protect our existing businesses." Or "We will operate each business on a self-sustaining basis; each will have to provide its own cash as well as make a contribution to corporate overhead." These rules are established either in a period of relaxed competition or out of political necessity.

The fundamental dilemma is that it always appears to be more economic to protect the old business than to feed the new one at least until competitors pursuing the new approach get the upper hand. Conventional financial theory has no practical way to take account of the opportunity cost of not investing in the new technology. If it did, the decision to invest in the present technology would often be reversed.

In military strategy the conventional wisdom for ground-based warfare is that the defender has the advantage. He is up on the hill with a clear view and can see the enemy coming. Military strategists from Von Clausewitz to Liddell Hart, as well as practitioners like Eisenhower and Rommel, all felt that if you wanted to take a hill held by a defender you needed to show up with three times his number or risk getting slaughtered. The spirit of that point of view is carried forward into business as well, with the belief that the defender, the competitor with the largest market share, the most knowledge of production processes and distribution, will have the advantage in combat in the marketplace. I believe the reverse is true. The defender is at an inherent disadvantage. He may not even know he is being attacked until the attack is well along. The attacker can hide in a niche. He is often more powerful than he appears, and more motivated.

A few years ago we polled the R&D vice presidents of 250 of America's largest firms who constitute the Industrial Research Institute. One of the most important findings was their belief that on average U.S. companies could double their R&D productivity. Such is the untapped potential of defenders that half the gain, they believed, could come from a more effective choice of projects and the other half from improvements in the work performed.

Our work at McKinsey corroborates this. When we analyze the technical spending of large companies, it is not uncommon to find 80 percent of the effort going to the defense of products that are more important for what they have contributed in the past than for what they are going to contribute to the future. This consumes funds that could be spent on the technical or market exploration of fields with higher potential or higher productivity. If there is usually a 5-to-1 difference in productivity between investments in emerging and mature technologies, then shifting just one dollar in five currently invested in the mature technology into the newer one would almost double the results.

In the 1K random-access memory, which was the first built, the productivity differences between the emergent and adolescent stages were on the order of 19 to 1. At McKinsey we have seen differences in productivity in electronics technology on the order of 20 to 1, even 30 to 1. That is 3,000 percent difference because of the basic choice of technology. People wondered how in the late 1950s tiny Texas Instruments could compete with giants like Westinghouse or Sylvania. Try multiplying TI's size back then by 30. By being on the right technological S-curve, TI compensated for much of its size and market power disadvantage.

In pharmaceuticals today, barely 10 percent of the R&D funding goes toward the newer biological approach to creating drugs. Yet my guess is that at least half the total innovations will come out of that area, so much higher is its potential.

Managers often talk about improving the productivity of a plant or of sales, but it is not often they expect anything higher than a 10 percent or 15 percent gain. We are talking 100 percent and 500 percent differences in technical productivity between competitors because one made the right technological choice and the other did not. No other area of management can match technology when it comes to potential gains in worker output.

Even if a defender succeeds in managing his own S-curve better, chances are he will not be able to raise his efficiency by more than, say, 50 percent. Not much use against an attacker whose productivity might be climbing ten times faster because he has chosen a different S-curve. All too frequently the defender believes his productivity is actually higher than his attacker's and ignores what the attacker potentially may have to offer the customer. Defenders and attackers often have a different perspective when it comes to judging productivity. For the attacker, productivity is the improvement in performance of his new product over his old product divided by the effort he puts into developing the new product. If his technology is beginning to approach the steep part of its S-curve, this could be a big number. The defender, however, observes the productivity through the eyes of the market, which may still be treating the new product as not much more than a curiosity. So in his eyes the attacker's productivity is quite low. We've seen this happen time and again in the electronics industry. Products such as microwaves, audio cassettes and floppy discs failed at first to meet customer standards, but then, almost overnight, they set new high-quality standards and stormed the market.

Even if the defender admits that the attacker's product may have an edge, he is likely to say it is too small to matter. Since the first version of a wholly new product is frequently just marginally better than the existing product, the defender often thinks the attacker's productivity is lower, not higher than his own. The danger comes in using this erroneous perception to figure out what is going to happen next. Too often defenders err by thinking that the attacker's second generation new product will require enormous resources and result in little progress. We know differently. We know from the mathematics of adolescent S-curves that once the first crack appears in the market dam, the flood cannot be far behind. And further, it won't cost nearly as much since the first product has absorbed much of the start-up costs. No doubt this will be a big shock to the defender who will tell the stock market analysts, "Well, the attacker was just lucky. There was nothing in his record to suggest he could have pulled this thing off." All true. From the defender's viewpoint there was nothing in the attacker's record to suggest that a change was coming. But the underlying forces were at work nevertheless, and in the end they appeared.

Reallocating resources is thus a painful business. At times

any decision that top management makes, any action they take may well be viewed as contrary to the company's best interests. Often CEOs will be roundly criticized by outsiders for venturing into new areas where they lack skills and for forsaking the tried and true. But in order to manage a technological discontinuity, that is exactly what they must do— forsake the past by abandoning a technology that, more often than not, has just entered the most productive phase of its S-curve. This dilemma captures the dimensions of the fourth era of managing technology which companies must now enter. It involves knowledge building, analysis, and the calculation of limits. And it involves, indeed requires, the conviction and courage to realize that sometimes it is necessary to cut off your arm.

FIVE

HOW LEADERS BECOME LOSERS

Businessmen go down with their businesses because they like the old way so well they cannot bring themselves to change. . . . Seldom does the cobbler take up with a new fangled way of soling shoes and seldom does the artisan willingly take up with new methods in his trade.

—Henry Ford
My Life and Work, 1922

Whenever technological discontinuities occur, companies' fortunes change dramatically. The leaders in the current technology rarely survive to become the leaders in the new technology. Their losses can vary from gentle to total, from embarrassment to humiliation. A discontinuity can result in the leader's retaining its number-one position in the marketplace but with a significantly diminished share. Or it can mean that the leader drops far down in the rankings or even withdraws from the business and goes into bankruptcy. This leaders-to-losers story has been played out not just by companies but by whole industries. And many times recently leading domestic industries have been losing to new international competition because of technological discontinuities.

Leaders lose their position because consumers prefer the new or improved, often cheaper products that result. Consider, for example, containers and packaging. Glass bottles, which helped Owens Illinois prosper, were replaced both by steel cans led by American Can and Continental Can and by paper cartons led by International Paper. Steel cans got their comeuppance in the beverage segment from the aluminum cans of Reynolds Metals and Alcoa. Glass bottles for soft-drink beverages have given way to plastic bottles, which meant new business for Eastman Kodak and Hoechst, the German chemical company. In the convenience stores, plastic milk jugs fill the refrigerator shelves that once were the preserve of plastic-coated paper cartons. Plastic pouches are now replacing metal frozen food packages as "Lean Cuisine" replaces TV dinners. Even tennis balls are not immune from packaging changes as they have moved from cardboard packages to metal cans and now to clear plastic containers. In all these cases one technology and its backer has ousted another in seemingly uninterrupted cycles.

I don't know of any comprehensive statistics that would stand up to academic scrutiny, but my feeling is that leadership changes hands in about seven out of ten cases when discontinuities strike. A change in technology may not be the number-one corporate killer, but it certainly is among the leading causes of corporate ill health.

ALLIED VERSUS BASF

Let's consider the devastating and almost identical results of discontinuities in four disparate industries—chemicals, tires, sugar and electronics. First, the case of a little-known but important chemical, phthalic anhydride (PA). PA is an organic chemical molecule, a building block in processes that result in paint thickeners and softer plastic luggage or auto upholstery and maybe stronger plastics in the future. So PA is an important industrial chemical now and may become more so in the future.

One way to make PA is to start with a raw material called naphthalene. Naphthalene has more carbon in it than is required to make PA, but lacks some oxygen that PA needs and which must be added. Given the chemistry we know how to do today, this limits the amount of PA it's possible to get out of a pound of naphthalene. Another chemical—orthoxylene—looks a lot like naphthalene except it has less carbon. It is possible to get more PA from orthoxylene than from naphthalene because all its carbon can be used. To be specific, whereas a manufacturer can get only 1.2 pounds of PA out of a pound of naphthalene, it can get 1.4 pounds of PA from each pound of orthoxylene. That is close to a 20 percent improvement and worth a lot in the chemical industry, where margins are often 10 or 15 percent. The fact that yields are higher with orthoxylene than with naphthalene does not necessarily mean that PA made with orthoxylene will be cheaper. If orthoxylene's price were more than 20 percent higher than naphthalene's, then the economic disadvantage would more than offset the technological advantage. And that indeed was the case until the early 1960s when more orthoxylene became available as a result of advances in oil refining. The greater availability relative to the demand for orthoxylene led to price decreases.

When the price of orthoxylene matched naphthalene, the advantage swung to orthoxylene. The S-curves (see Exhibit 13) for these approaches to producing PA show how orthoxylene's technological performance shot right through naphthalene's. Naphthalene was constrained by its lower technology potential. If prices for the two raw materials stayed the same, the only way naphthalene users could stay ahead of orthoxylene users was to pour more money into developing the manufacturing process. But ultimately, because of the physical limit imposed by naphthalene's excess carbon, they would lose.

The leader in naphthalene technology was Allied Chemical Corporation (now Allied Corporation), followed in the United States by Monsanto. The leader of the orthoxylene

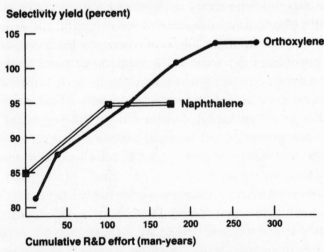

13   The S-Curves for Phthalic Anhydride and Orthoxylene.
The advantage once held by Allied in naphthalene technology was lost to orthoxylene, produced first by BASF and then by U.S. companies under license.

technology was West Germany's Badische Analine and Soda Fabrik (BASF), which has a sizable United States operation.

For a variety of reasons, Allied stayed with naphthalene and chose not to go into the new process, even though BASF had improved the technology sufficiently to challenge Allied for its market dominance. Allied couldn't compete when the price of orthoxylene began to decline. For its part, Monsanto did not stick with the old technology but wisely bought a license to the orthoxylene process from BASF, maintaining its number two position.

Why did BASF license Monsanto? Why didn't it try to keep all the profits for itself in the U.S.? Basically, BASF didn't want to add more capacity to an already well supplied U.S. market for fear that prices might tumble. Instead, BASF hoped to generate more profits by providing licenses to Monsanto and other producers who would replace their current naphthalene-based production of PA with the superior orthoxylene method. Part of BASF's strategy involved a two-step approach often used in the chemical industry. It developed a catalyst, which it then improved. The first generation of the catalyst went to its licensees, the second and improved generation into its own plants. Thus, Monsanto and the other licensees could compete effectively against the naphthalene producers, including Allied, but were still less effective than BASF.

Unfortunately for BASF, U.S. producers were too enthusiastic about the new process. They not only replaced existing naphthalene-based capacity with the orthoxylene process, but they then added new and unneeded capacity. Prices collapsed and, for a while, all producers lost money, the higher-cost naphthalene producers more than the orthoxylene producers.

Does an understanding of S-curves change our perspective on these events and improve our understanding of the competitive dynamics? Consider Exhibit 14 which shows that scientists working for Allied Chemical, Monsanto, Chevron and others spent some 100 man-years of effort between 1940 and 1958 seeking more efficient ways of making PA from naphthalene. During that period performance improved steadily. But from 1958 to 1972 scientists exerted an additional 70 man-years of effort and achieved only limited progress. The process had fallen victim to the rule of diminishing returns. Eventually progress on PA from naphthalene stopped completely. To anyone asking after the fact, "What did you get for what you spent?," the cold analytical answer would be, "Sixty percent of the money we spent produced significant results, but 40 percent—the last 40 percent—produced nothing. It was wasted!"

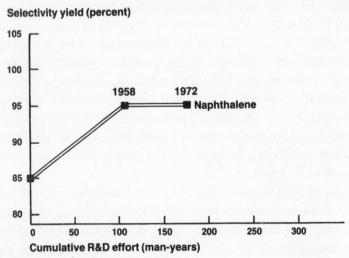

14 Naphthalene's S-Curve.
No progress was made after 1958 because the process was at its limit.

If Allied had been able to eliminate their unproductive effort, they could have almost doubled the technical progress made for the effort put in. Or said another way, they could have eliminated 40 percent of the total cost of improving the technology. That would have helped even if there were no other technologies to follow. At the least they could have taken that money as profit.

For naphthalene-based PA, the slope of its S-curve or productivity during the middle, adolescent stage was five times greater than productivity in the last, or mature stage. That's quite an advantage. Said another way, if you were a chemical company operating in the adolescent stage of PA development, you were getting five times the result a competitor would get if it were investing in the mature stage. Actually the estimate of "five times as productive" is a bit conservative, because Allied Chemical was pursuing the naphthalene route in the mature stage, while BASF was using orthoxylene in the adolescent stage. Given this situation, BASF, although initially it didn't get as much PA from its process, was making about 12 times as much progress per man-year spent as Allied. And therefore it was bound to catch up, and fast. The improvement from 5 to 12 comes from the learning done between sequential generations of products. S-curves frequently tend to get steeper as each new wave of technology comes in. This happens probably because each new generation builds on the accumulated knowledge base of the prior generation.

So a 12-to-1 economic advantage for the attacker in this case. Even that understates the advantage because the limits of the orthoxylene technology were about 20 percent higher than the limits of the naphthalene technology. In the graph in Exhibit 13, these S-curves cross each other. That is where BASF began to get more pounds of PA from its process per pound of raw material than Allied did. From then on BASF was unstoppable; Allied had reached the limits of its technology and could not ever hope to catch up with BASF no matter how much it invested in the old naphthalene technology.

_____ DU PONT VERSUS CELANESE

Let's look at S-curves at work in another product—tire cord. Its performance parameter is more complex than a simple ratio like pounds of product per pound of raw material. It involved such things as the cord strength, heat stability, adhesion and fatigue. They combine to give the tire the properties consumers want—a smooth ride, endurance, blowout protection and low cost.

A performance parameter has to meet two conditions. It

has to be something of value to the customer, and it has to be expressed in terms that make sense to the scientists and engineers in the company. The combined overall performance parameter for tire cords meets these conditions because each of its component parts meets them. The overall measure is the result of combining the component parts after weighting them in terms of their importance to customers. With this integrated performance parameter, it's then possible to construct an S-curve.

In this case, all performance factors are related to the best performance of cotton since it was the first tire-cord fiber. It has arbitrarily been assigned a value of one. The first synthetic tire cord was rayon. Stronger than cotton, it allowed thinner tires to be made. And rayon did not rot like cotton, so tires made from it lasted longer. During the course of several years (Exhibit 15), over $100 million went into the

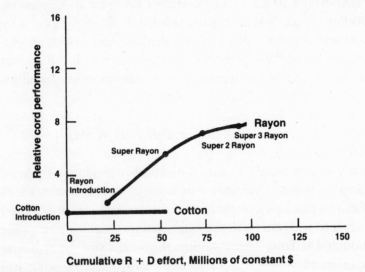

15 The Improvement of Rayon.
A significant portion of the $40 million spent after 1962 might have been saved if limits had been understood.

improvement of rayon. But the money spent produced differing amounts of improvement in performance. The first $60 million brought an 800 percent gain over where it began, the next $15 million a 25 percent improvement, and in the early sixties the final $25 million brought only a 5 percent improvement as rayon technology reached its limits. As with PA, if the companies involved had known about checking for limits, their investment strategies might have been different. At a minimum they might have saved a significant portion of the $40 million spent after 1962.

Initially, the leading rayon producers were American Viscose and Du Pont. After World War II, Du Pont switched from rayon to its proprietary nylon tire cord. Over the years Du Pont and American Viscose tried to better the other's product with a series of improvements. Because nylon had higher limits, rayon began losing share. Finally, American Viscose faded away and was bought out by FMC. But for Du Pont it was a short-lived victory.

Nylon tires weren't without their problems. Drivers suffered from what became known as "flat spots." On cold days in winter, the nylon tires froze when the car was parked for a while, creating a flat spot on the bottom of the tire. As the spot revolved it caused the tire to bump against the ground. Before long Detroit's engineers were putting pressure on the tire manufacturers to get rid of the bumps. In turn the tire manufacturers encouraged their suppliers to find alternatives to nylon. One alternative was polyester—the material that was also used for double-knit suits. It was made by Du Pont, Celanese and others. In the end Celanese won this battle even though Du Pont had a strong overall position in polyester (Exhibit 16).

Du Pont embarked on a dual approach. It pursued substitute polymers like polyester but also tried to improve ny-

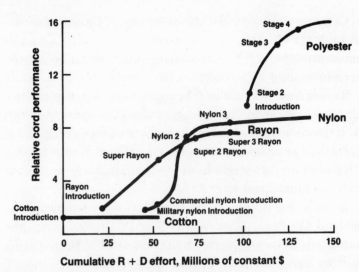

16   From Cotton to Rayon to Nylon to Polyester.
Du Pont, not understanding where nylon was on its S-curve, got little for its last $75 million of R&D, while Celanese progressed much faster with less money because polyester was just starting its curve.

lon's flexibility to get rid of flat spots. Unfortunately Du Pont didn't know where nylon was on its S-curve. That lack of information proved costly. Nylon was closer to the limits of its technological potential than anyone had guessed. Lots of money poured into R&D could not, in fact did not, make much of a difference. But that was not the case for polyester. Polyester was still in its adolescence. Celanese's adolescent polyester technology had a 5-to-1 advantage over Du Pont's mature nylon technology. As with PA, there was an increase in the steepness of the S-curve from one generation of technology to the next, and polyester had much higher limits. The polyester limits were about 16 on the performance chart versus 8 for nylon. Once perfected, the polyester corded tire would last longer and stay more flexible at lower temperatures—no flat spots—than the very best nylon-corded tire.

Celanese could spend half as much as Du Pont on tire cord R&D, but still progress two-and-a-half times faster than Du Pont since its productivity was five times higher. Less money, more progress.

The difference between these two competitors was essentially one of technical choice, and technology more than any other single variable dictated the relative effectiveness of Du Pont and Celanese.

In pursuing substitute polymers, Du Pont concentrated its efforts in two areas—polyester and Kevlar, an extremely tough fiber. Kevlar was a great product but extremely difficult to make. It was still in the early stage of its S-curve. So this left polyester and here Du Pont should have had an edge. Its manufacturing technology was already in hand. Du Pont was both the leading maker of polyester and the leading producer of tire cord. So you might think Du Pont inevitably would become the country's leading producer of polyester tire cords. But it didn't. Celanese did. Why?

Part of the problem at least can be traced to how Du Pont organized itself internally. When nylon became a significant product for Du Pont it made it an independent profit center. The aim of its Nylon Department was to make as big a return on Du Pont's investment in nylon as it could. To keep ahead of outside competition, the Nylon Department had spent much time and effort to create a tire-cord development center. There the Nylon Department's engineers tested the latest developments in tires, enabling them, in the best corporate tradition, to stay close to their customers, the tire manufacturers. So when the Polyester Department decided it wanted to break into the tire business, it seemed to make sense for the tire-cord development center to test polyester tire cords too. The expensive alternative would have been for Du Pont to build and operate another tire-cord development lab within the polyester department.

Nobody outside Du Pont knows the full story, but from what I've been able to piece together, it goes something like this. The Polyester Department brought its tire for testing to the Nylon Department and was told, "Gee, this is a terrific product. It's almost as good as the latest nylon tire cord. It just needs a bit more development." For the polyester designers that was good news. They enthusiastically continued working away for a year or so improving their product design. When they returned for more testing they were told their product was absolutely fantastic and superior to anything made of nylon. Unfortunately, Du Pont had just approved a new investment in a nylon tire-cord facility and would have all the necessary tire-cord capacity it needed for some time. All that top management could promise the Polyester Department was that when the nylon capacity was used up, Du Pont would invest funds in polyester tire cords.

So for several years, in order to protect its investment, Du Pont pushed nylon, despite the fact that its leading customer (Goodyear) had publicly asserted that polyester made better tire cords.

Celanese was also big in the fiber business, but it didn't have a position in nylon to protect. Celanese had its own tire development lab, which quickly turned out a polyester tire cord that was snapped up by the tire manufacturers. The net result was that Du Pont lost a precious opportunity to establish a preemptive position in the marketplace. Celanese leapt forward with a commanding position. In five years, in the latter part of the 1960s, sales of tire cords grew only slightly, but Celanese captured over 75 percent of the market.

Du Pont didn't lose out because it was unaware of the

technical possibilities of polyester, but because it implicitly assumed that, as a corporation, it could control the pace of innovation in the marketplace. It couldn't. Only a monopoly can do that, and Du Pont didn't have a monopoly.

The story doesn't quite end there. What we've been examining is a game within a game. As these competitive battles were taking place in bias-ply tires, the newer radial tire was making great headway in Europe because it had a much longer life and was better-handling than bias-ply tires. Michelin, then a relatively unknown competitor in the U.S., was preparing for an attack on the United States market. Through the late 1970s, Michelin was the most successful new competitor to break into the U.S. market in this century and it did so by riding a whole new technology in the form of the radial tire. Michelin's success with radials was in large measure gained because American tire makers were late to switch to radials. They stayed on the wrong S-curve.

Michelin had something else going for it—changing performance parameters. Customers wanted long tire life and radials gave it to them. True radials had a rougher ride, but many drivers liked this "sports car" feel. Initial sales of radials were to sports car enthusiasts who bought them as replacement tires at Sears, which had an exclusive distribution arrangement with Michelin. But this niche was a precursor of a much larger market shift. U.S. manufacturers were wedded to their bias-ply technology, however, and failed to see the niche as anything more than that, a specialty segment. It was anything but that.

As *Fortune* magazine said in 1974, ". . . the radial cost more than conventional tires and was not immediately recognized as the tire of the future. Akron, in particular, tended to sniff at the belted radial as a 'European' tire—okay perhaps for little doodlebugs scooting around on cobblestones, but not suited to the high-powered sofas on wheels that Americans piloted."*

If you think the game is over, think again. The competitive game never ends. Du Pont is back in there pitching hard ball with Kevlar, which really looks to be superior to all other tire cords—at a price—for radials. Du Pont is the only company with Kevlar, so they are in a very strong position. But there may be yet another game in town—LIM tires. It stands for Liquid Injection Molded tires. If that doesn't help you much, think of them as an entirely new tire, not made out of rubber but made out of plastic. A new S-curve. LIM tires have less tread abrasion, enable better gas mileage, are light and handle better than radials. Who makes them? A little Austrian company named Lim Kunstoff Technologie in Kittsee, Austria. Their primary business is not tires at all; it's building plastic-shoe-manufacturing equipment. What is their chance for success? Michelin says not much in private meetings with rubber suppliers. Then again 100 percent of Michelin's assets are tied up in radial-tire-making equipment. A 1983 survey rated the LIM KT tire superior to Bridgestone's (a Japanese radial tire manufacturer). Both German and British car manufacturers are testing them. A British automotive writer, Stuart Marshall, evaluated them and said, "I would gladly have a set of LIM tires on my own car."† What will happen? Who knows. In 1984, Lim Kunstoff produced 15,000 commercial tires. It is shooting for a million in 1985. Will people really buy them? They are more expensive than radials, but people bought radials when they were more

---

* "The Michelin Man Rolls On to Akron's Backyard," *Fortune,* December 1974, p. 138.
† Marshall, Stuart, "First Test Report on the Revolutionary Plastic Tire," *Popular Science,* April 1983, pp. 100–102.

expensive because on a total cost basis they were cheaper. Maybe that will happen here too. Maybe it won't. By the way, the plastic tire has Kevlar tire cords.

WELCOME TO GERMANIUM GULCH

Discontinuities happen in electronics too. If they didn't, we would probably be talking about Germanium Gulch today and not Silicon Valley. Few people remember that in the beginning, and the beginning was in 1952, silicon was not the stuff of which transistors were made. Germanium was. It is a basic chemical element, mined with copper, zinc and lead ores.

In every atom there is an energy band gap. It is the amount of energy that must be put into an atom to knock off an electron. That's important because if it wasn't possible to knock an electron off an atom, materials would not conduct electricity, which is basically the flow of electrons between atoms. In fact, there are many elements, like carbon, with which this is rather tough to do. There are others, like copper, with which it is easy to do. In these materials electricity flows with ease. Some elements, like silicon or germanium, fall between these two extremes. It is neither particularly hard nor particularly simple to knock their electrons off. These elements are called "semi" conductors because they conduct a little bit of electricity but not a lot. These materials are not used in power cables, because they cannot carry much current. But as the world now well knows, semiconductor chips are essential for carrying little bits of current in a computer in order to add two numbers together, or to carry a radio signal or a telephone call.

One way to characterize semiconductors is to rank them

by their band gaps—by how easy it is to knock an electron off the atom. A small band gap means it's rather easy to do. Germanium has a rather small band gap, which is why the people at Bell Laboratories who invented the transistor used germanium first. Alas, the ease of knocking electrons off is a double-edged sword. It also makes it easier for unwanted impurities in the manufacturing process to degrade the performance of the final product. Thus it was hard to produce germanium devices reliably. High reject rates when germanium chips were manufactured drove their costs up.

Enter silicon, with a higher band gap but greater reliability. This reliability is of tremendous commercial consequence because it means that products need to be serviced or replaced less frequently. High reliability translated into low cost, and therefore into a competitive advantage, at least over germanium, for the producer using silicon-based technology. Thus, for fundamental physical reasons, silicon producers such as Texas Instruments and Motorola took the market away from germanium producers like Hughes and Sylvania. Another case of a technological discontinuity precipitating a competitive rearrangement of the leaders, but this time between competing new technologies. There was nothing that germanium producers could do given the lower limits of their process. It was naphthalene and orthoxylene all over again, but with even bigger differences in performance. The result was really inevitable. Yet for many people it was a tremendous shock to see giants like Hughes or Clevite overtaken by what was then a tiny company like Texas Instruments.

The idea of limits and S-curves applies to low-tech businesses as well. Most people regard the sugar industry as a low-tech business, but these days there's more to sugar than was once the case. "Sugar" is not one thing, but rather a whole family of chemicals with reasonably familiar names such as sucrose, fructose, glucose, galactose and mannose. The sugar on our breakfast table is sucrose. It comes from sugar cane. Sucrose is not really a simple sugar but a chemical combination of two simple sugars, glucose and fructose. This technical detail has huge commercial implications.

For many years soft-drink makers put costly sucrose made from relatively rare (in the United States) cane in their products. But by the time these were packed, shipped, stored, displayed and sold, most of the sucrose had undergone a natural transformation into its component parts, glucose and fructose, which—coming from more abundant corn—were cheaper. Consumers, of course, didn't notice, but the manufacturers increasingly became concerned that they were wasting money, particularly since fructose itself can be simply made from glucose through a straightforward chemical transformation. Eventually the corn sugar processors began to ask, "Why can't we convert the low-priced glucose into a mixture of glucose and fructose, which will compete against sucrose but have a lower price? After all, the glucose-fructose mixture is the same as the mixture the consumer actually drinks." The answer was they could, if they could find an economic way to convert the glucose into fructose. Their chemists soon showed them how.

Before long the corn processors had a product that tested well in soft drinks and had a lower cost than sucrose. After agonizing over whether the consumer would think that fructose and glucose tasted the same as sucrose, the soft-drink manufacturers began to replace sucrose with the glucose-fructose mixture—the so-called High Fructose Corn Syrup (HFCS). The net result: a shift in leadership in the sugar market from Gulf & Western, the leader in the sucrose business, to A. E. Staley, the leading producer of HFCS. So technology change can unseat leaders even in supposedly "low-tech" industries.

_____ THE DEFENDER'S PARADOX

Technological transitions lead to the demise not only of individual product lines but of whole industries. In the mid-1950s, which marked the commercial start of the modern electronics era, vacuum tubes were about a $700 million market. The transistor market at about $7 million was tiny by comparison. As Exhibit 17 shows, during the quarter of a century between 1955 and 1982 there was almost a complete turnover in industry leadership. Only RCA and North American Philips (through its Amperex subsidiary) were successful producers of vacuum tubes who became successful producers of transistors and integrated circuits. They were the only ones that survived this technological discontinuity. The companies on the list are "merchant" producers of electronic components. That is, they produce these devices for others to buy, not for their own use. If we added "captive" suppliers, those companies that produce solely for their own use, we would have to include IBM and Western Electric, the manufacturing arm of the old Bell System. If we did that, we would find that four companies made it through the discontinuity and that two of them, IBM and Western, are industry leaders.

The chart also shows the entrance of the Japanese man-

	1955 (Vacuum tubes)	1955 (Transistor)	1960 (Semi-conductor)	1965 (Semi-conductor)	1970 (Semi-conductor)	1975 (IC)	1980 (LSI)	1982 (VLSI)
1	RCA	Hughes	TI	TI	TI	TI	TI	Motorola
2	Sylvania	Transitron	Transitron	Fairchild	Motorola	Fairchild	Motorola	TI
3	GE	Philco	Philco	Motorola	Fairchild	National	National	NEC
4	Raytheon	Sylvania	GE	GI	RCA	Intel	Intel	Hitachi
5	Westing-house	TI	RCA	GE	GE	Motorola	NEC	National
6	Amperex	GE	Motorola	RCA	National	Rockwell	Fairchild	Toshiba
7	National Video	RCA	Clevite	Sprague	GI	GI	Hitachi	Intel
8	Rawland	Westing-house	Fairchild	Philco/Ford	Corning	RCA	Signetics	Philips
9	Eimac	Motorola	Hughes	Transitron	Westinghouse	Philips	Mostek	Fujitsu
10	Lansdale Tube	Clevite	Sylvania	Raytheon	American Micro	American Micro	Toshiba	Fairchild

17 **From Vacuum Tubes to Semiconductors.**
Technological transitions lead not only to the disappearance of individual product lines but to the demise of whole industries.

ufacturers (NEC, Fujitsu, Toshiba, Hitachi) into the market in the early 1980s. These entrants have taken a substantial share of our U.S. market and have affected the U.S. balance of trade.

Four of the original ten vacuum tube leaders—National Video, Rawland, Eimac and Lansdale Tube—were makers of receiving tubes such as TV or radio tubes. They chose not to enter the solid-state business, perhaps feeling they could not compete effectively. Of the other six, Westinghouse dropped out of the merchant market by 1960, Sylvania sold its assets by 1965 and shortly after 1970, GE called it quits. By 1975, RCA and Philips (through its acquisition of Signetics, not through the success of Amperex, which it owned in 1955) were the only vacuum tube makers still on the solid state list.

There are three variants of error in these case histories. First is the decision not to invest in the new technology. National Video, Rawland, Eimac and Lansdale Tube are examples. The second variant is deciding to invest but picking the wrong technology. Hughes, Transitron and Clevite all chose to get into solid-state electronics with germanium and lost.

The third variant is cultural. Companies failed because of their inability to play two games at once: to be both effective defenders of what quickly became old technologies and effective attackers with new technologies. In 1968, fully five years after it had lost its shootout to Du Pont and nylon, American Viscose was still producing rayon even though rayon's market share had fallen from almost 50 percent to barely 20 percent. *Tire* magazine wrote, "Chester Dodd, the Akron representative and Jim Curly, head of tire cord development of American Viscose, point out that their company makes both rayon and polyester. And they still sincerely believe that the best tire cord is rayon." What those executives didn't tell the magazine was that American Viscose was producing negligible quantities of polyester. They had great faith in their old product, but it probably cost American Viscose many million dollars in lost profits and unneeded capital expenditures.

Like so many other companies, American Viscose faced the defender's paradox. It couldn't overcome the instinctive tendency of protecting its existing business even though each investment produced ever diminishing returns. As a result it did too little, too late. Westinghouse did this, in part, by burying its electronics operation so far down in its organization structure that top management had scant chance to take any notice of it, much less give it support. Du Pont did the same with polyester.

What we see, particularly in this last case, is that technological change necessitates significant organizational change. Changes in leadership and career paths. New winners and losers. Disruption of prior expectations. Again and again companies lose their leadership not only because of weak strategies but also because of strong cultures.

SIX

THE DEFENDER'S DILEMMA

Ignore, ridicule, attack, copy, steal.

—Arthur Jones, inventor of Nautilus exercise equipment, listing the reactions of competitors, 1985

*I*t is relatively easy to spot new technologies on the horizon and to decide to monitor them or perhaps invest in them. What is much harder, indeed agonizing at times, is to stunt the growth of the older technology by withholding development funds from it even though progress can be made. People lose their jobs, friendships are destroyed, often the entire business must change. These are facts we don't want to face.

In 1967 then Vice President for Finance of NCR, J. J. Hangen, told the press, "The base of NCR's revenues comes from cash registers and accounting machines. Computers both support and *protect* NCR's traditional product lines." Hangen and NCR clearly did not believe that computers and electronic technology were about to revolutionize its traditional line of business. In fact, Hangen was making a virtue out of the company's sticking to its knitting, of doing what it knew best. The press urged them on by writing, "The company has branched out into computers but at great cost. Its competitors and customers have been puzzled by a slow pace in product development on its home grounds, retail systems, and banking."*

* "The Rebuilding Job at National Cash Register," *Business Week*, May 26, 1973, pp. 82–86.

By 1970, NCR said in its annual report, "NCR has continued to stress the further evolution of its wide line of basic business machines." Time was passing, but NCR was not paying attention to what was happening in the marketplace. As one magazine noted, "In 1971, NCR was still clinging to the past, endlessly refining an obsolete electromechanical technology, even though the computer revolution was on the verge of overwhelming it."* In 1971 a little company called DTS brought out the first electronic cash register. The transition was unfolding.

All this time, NCR was dabbling in computers. Trying to use them to break into newer lines of business, while leaving their traditional cash registers tied to the electromechanical process. How NCR thought it could compete against IBM in computers is hard to understand, particularly since IBM already had introduced its 360 model and IBM's budget for R&D was a sizable fraction of NCR's total sales. What made things even trickier for NCR was that during the early 1970s the market for cash registers went through some turbulent times as customers, worried both by the recession and the uncertainty of electronic versus electromagnetic machines, delayed purchases. The net result of all this activity was that electromechanical cash registers went from 90 percent of the market in 1972 to 10 percent in 1976 (Exhibit 18).

For NCR the experience was traumatic. The company was forced to write off $140 million of electromechanical equipment barely one year after making it. The next year they had a loss of $60 million. The board of directors ousted the chairman three years before his contract was due to end. The new chairman then put 28 out of 35 corporate officers on "consultant" status for the rest of their careers. Roughly

* Seneker, Harold, "Leading NCR out of the Wilderness," *Fortune*, February 18, 1980, p. 74.

DELIVERIES OF NEW CASH REGISTERS IN THE U.S.
Market Share

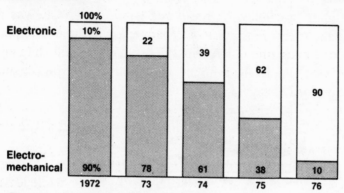

18 Discontinuity in the Cash Register Industry.
In four years, 80 percent of the market for cash registers was lost to manufacturers of electronic products.

20,000 workers in the electromechanical plants were let go. A sad event, but necessary to avoid bankruptcy and lose the remaining 70,000 jobs. Stockholders did not make out well either. They watched the value of their stock in NCR fall to about 10 percent of what an equivalent purchase of competitor Burrough's stock bought in 1964 would have been worth (Exhibit 19). Some of that lost ground has now been regained. But much of this trauma could have been avoided if NCR moved into the new technology sooner.

RCA is another case of how difficult it is to change technologies. It was, by far, the most successful of the leading tube makers in crossing the discontinuity to solid state, but even RCA was plagued by the difficult choices of which technology to back. It had to face questions like "Why should we cannibalize our profitable tube business for uncertain profits from a rapidly changing, solid state business?" These questions led to vacillation.

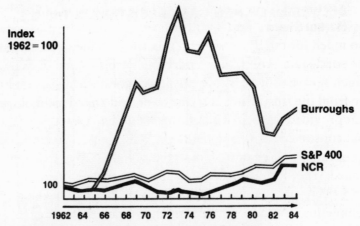

19 Stock Performance Reflects Technological Strength.
Stockholders in NCR watched as their company struggled to regain what
it had lost to Burroughs, the leading producer of electronic registers.

From the outside it looked like RCA was firm in its decision to move ahead into solid-state electronics. It organized
a group to develop solid-state devices, primarily for hearing
aids, transistor radios and military applications. But this
group reported to the head of the electronics business, who
was also in charge of vacuum tubes. Naturally, he tried to be
pragmatic and protect the cash flow of the on-going tube
businesses. It made no sense to cannibalize a proven source
of income. Smaller competitors like TI, however, had no such
cannibalization to worry about and so moved ahead with
more direction and purpose.

Eventually RCA's managers saw they had a problem and
changed the reporting relationships. No longer would the
solid-state group report to the vacuum tube executive. But
where would it report? RCA didn't want to leave it hanging
off in space somewhere, so had the group report to a senior
executive on the head office staff.

That worked well until the red ink started to flow. The losses, quick pace and wholly strange technical ideas were too much for the RCA general executives, so they transferred the solid-state group back under the vacuum tube division, which now had new management. All these changes, resulting from the frustration of trying to defend an old technology that provided the cash while attacking with a new technology that consumed cash, took their toll. With each reorganization the strategic direction of the group was changed. Morale waned. The shuffling undermined the engineers' self-esteem. They were always being asked to stop things just short of completion. Always being rushed to start something new without adequate preparation.

All of this was taking place at the same time RCA was aggressively expanding into frozen foods and books and trying to beat IBM at computers. Not surprisingly, RCA's semiconductor fortunes slipped from fourth place in 1970 to eighth place in 1975. To keep this all in perspective, it was also during this time that Motorola slipped from second to fifth and GE and Westinghouse dropped out. But with all their troubles, RCA did relatively well.

_____ CULTURE SHOCK

The purpose of these stories is not to berate former management of these companies but rather to illustrate that changing the underlying technology of a company or a division requires changing its culture. It is one of the most formidable tasks management can face.

In order for companies to survive a discontinuity they must, as *Harvard Business Review* editor Alan M. Kantrow once put it, "face the rather unpalatable reality that there may have to be fundamental changes in who they are, what they do and how they do it, as wrenching and as dislocating as it may be." In a very real sense they will have to undergo a metamorphosis.

Technology is an integral part of the corporate culture. In fact, it underpins it. Changing technology is as difficult as changing a company's culture. One of the hot new industries, biotechnology, shows how difficult it can be. Everybody expects the big chemical companies to lead the way in biotechnology. But maybe they are mistaken. Maybe the chemical companies will be just like the vacuum-tube manufacturers.

Fundamentally the chemical industry and the "biochemical" industry are quite different. In the traditional chemical industry the rule of thumb is the bigger the better when designing a plant. A large part of the cost of a plant is metal, mostly in the form of steel spheres. As you increase the diameter of the sphere, the area contained increases as a function of the diameter squared, but the volume increases as a function of the diameter cubed. Thus if the diameter is doubled the surface area is increased by two squared or four, and volume by a factor of eight. Since the cost of the sphere is a function of its surface area and not its volume, the cost per unit volume has been halved.

Traditional chemical plants are not only huge, they are custom built. They need engineers knowledgeable in handling reactions at temperatures up to 1000 degrees Fahrenheit, and who know, for example, how to put naphthalene in one end and get phthalic anhydride out at the other. These engineers have become very skilled at what they need to know. How to handle high pressures (sometimes up to hundreds of times atmospheric pressure). How to carry out and control complex steps in huge vessels, and indeed how to build and transport these leviathan-like vessels from the fabrication sites to plant sites.

What they've never had to do, and therefore are not necessarily good at, is design small pieces of equipment that don't require much energy or pressure but do require biological purity (no microbes floating around), and very gentle handling of the fluids inside them. These are the requirements for a biochemical facility, where the bugs do the work. In biochemical plants bigger does not necessarily mean better, because it's much harder to handle large volumes of bugs than small ones. And the bugs do their work at low temperatures; they die at high temperatures as they do if you slap them around inside the conversion pots the way you do inanimate chemicals.

So the tricks of the trade that made an engineer unusually productive in the chemical business aren't necessarily going to sustain him in biochemicals. He has spent a career learning how to save energy while handling reactions at high temperatures, but in the future fewer managements may care because that may not be the key. The ability to mix a vat of stuff with a big paddle just isn't going to work in the biochemical industry because the big paddle may beat the microscopic organisms to death. What is now needed is knowledge of process controls and biology. The engineers must invent new ways of stirring or find themselves out of work and their companies out of business.

Suddenly the chemical business could start to look much more like the pharmaceutical industry. It is reminiscent of the transition from vacuum tubes to solid state where the designers knew about electron flows and tube characteristics rather than the N-, P-, and C-channels of the new technology. The presumption that an engineer who was a great designer of chemical plants will be a great designer of biochemical plants doesn't hold up. Too many critical factors have changed. It's like assuming a great football player is going to turn into a great rugby player, or a great soccer player. Although the games look similar, the specific skills that account for excellence are very different. Different muscles, different techniques, different degrees of endurance.

The major cultural difficulty in managing through technological discontinuities is making skill transitions. What any company amounts to, no matter how large its assets base, is the skill of its people. If that skill base is rendered useless, if other skills turn out to be more relevant, then management faces a very difficult problem. It must direct all its own skills at anticipating this problem because it takes so long to fix. You just can't change people's skills overnight.

Then there are strategy changes as well. For instance, consider what might happen if the biochemical business were to take over from the chemical industry. We could easily find that small plants were more economic than large plants and that they could be located all around the country at customers' sites. Distribution costs would be much lower. If these developments occur, what will happen to the big centralized plants on the Gulf Coast and all the employees in that area? It could mean hard times.

We have seen this kind of change in strategy before. Air Products accomplished it after World War II. With a limited amount of capital, founder Leonard Pool set out to produce industrial gases in a new way. His approach was to build small air liquification plants to provide oxygen and nitrogen at the sites of his customers, the steel mills. His manufacturing costs were higher than those of his competitors with large centralized plants, but he gained two advantages. He didn't have to pay any costs for transporting the liquified gases, and he picked up some valuable contracts from his customers who guaranteed to take or pay for his output. Pool then used these contracts as collateral at the bank, avoiding the need to raise equity funding and dilute his ownership. Pool added one more wrinkle. He built his plants 20 percent larger than what his contract called for, and proceeded to sell the extra capacity to local, smaller customers at a great distribution cost advantage over his competitors. On the basis of this strategy of decentralized plants, Air Products grew fast. It now has sales of over a billion dollars and is one of the more profitable companies in its industry.

A decentralized business doesn't require the kind of functional organizations that chemical companies currently have. Systems can also be looser. Capital expenditures in the chemical business are centralized because of the size and costliness of the plants. But in a biochemical business with its contracts and smaller plants, corporate staff need not be involved.

There is no reason why the ideas and strategies developed by Air Products cannot work in the biochemical industry if the technology develops as it might. The bigger question is "Will a chemical company which is used to thinking in terms of large centralized facilities be inclined to undertake these kinds of changes?" It's very unlikely.

THE EXPERIENCE OF JAPAN

No wonder, then, that companies have such difficulty switching to a new technological base. It means changing everything. But it has been done. The Japanese jumped into the solid-state business in the early 1950s. As in the United States, their group of original transistor-makers had representation from both established receiving-tube manufacturers (Toshiba, Matsushita, Hitachi, NEC, and Mitsubishi) and from new entrants (Sony, Sanyo, and Fujitsu). Unlike the U.S. experience, however, by 1980 Toshiba, Matsushita, Hitachi, NEC, and Mitsubishi remained leading manufacturers. They had all made the leap from germanium into silicon, from transistors into integrated circuitry, from analog into digital, from integrated circuits into microprocessors.

Why? What accounts for these differences? The reasons are both economic and social. In the early 1950s the Japanese electronics industry was fairly small and lacked an indigenous technical base. Needing technology and looking for a way to ultimately crack export markets, the Japanese went looking for technology and found it in the United States. They began to produce both transistors and transistor products like the pocket radio. Because of the low labor rates and yen value at the time, the Japanese products were much cheaper to make than those made in the United States. As a result the Japanese found a ready market for their products. The industry grew and prospered, stimulated further by intense internal competition.

While established tube-manufacturers in the United States perhaps viewed the new technology as a mixed blessing (after all, it raised the issue of whether they should cannibalize their own products), there was no such dilemma in Japan. There was little to protect. The assets invested in tube production were miniscule. The opportunity for getting a jump on foreign competition in export markets was immense. The tube manufacturers moved quickly.

Some new competitors entered the Japanese market. Sony, for example. But there was not a ready supply of entrepreneurs, nourished by a willing venture capital community, to provide for real competition with the majors. The net result was that the Japanese receiving-tube companies had 79 percent share of their semiconductor market in 1959 and still had it in 1979.

Certainly the business environment in which the Japanese achieved their successful crossing of the electronics discontinuity was rather special from an American perspective. Few of us would expect these conditions to be duplicated for American companies. This isn't to say that these Japanese companies did not have to wrestle with change. But the circumstances were different in Japan and the U.S. The Japanese had little to protect, a great deal to gain and a willing source of supply of the missing ingredient—technology. They moved with dispatch and the rest is history. The message is that if the incentives are there, and are seen to be there, companies can manage through transitions; but it requires starting early, because skills and culture have to change. Starting early presumes a felt need to change, to move in a new direction, to abandon the old and to get on with the new. The Japanese majors had that incentive more clearly than the American majors. They were much more like Texas Instruments and Motorola than they were like GE, Sylvania and Raytheon.

RESPONDING IN TIME

Anticipation is the key. A company has to start early. The less time it has, the more money it will have to spend to accomplish the transition. At some point, the amount of money needed will exceed resources. That may mean bankruptcy or significant loss of position. That's what happened at Addressograph Multigraph in the late 1970s when its former chairman Roy Ash tried to move the company out of electromechanical technology and into electronics almost overnight. The result was bankruptcy. Companies cannot make a cultural change that fast. The managers make mistakes, and the company suffers.

The better companies get ready for change early. A good example is Monsanto. For decades Monsanto made its money in commodity chemicals. With the price of oil rising in the 1970s, Monsanto began to face new competition in petrochemicals from the oil companies. Shrewdly, Monsanto began turning its back on businesses in which it had over $1 billion invested and moved instead into more promising areas like biotechnology. In 1975, for example, it gave an $8 million contract to Harvard Medical School. As its reputation grew, Monsanto hired Howard Schneiderman, who was formerly head of the biology department at the University of California, Irvine, to head up the biological effort. The whole operation reported to then Vice Chairman, now Chairman Louis Fernandez. Then came small venture investments and acquisitions over a period of time. Monsanto did not try to mix the old and the new skills, although some managers did transfer, and they didn't try to make the switch overnight but started the transition early and slowly. Now the group is set up basically as an independent, stand-alone business.

Today, these investments are beginning to pay off for Monsanto. In September 1985, Harvard announced the discovery of angiogenin, a substance (as well as the gene that produces it) that "initiates the growth of any human organ. The first practical result from the discovery may be a new strategy against cancer." *

But Monsanto and the Japanese tube manufacturers are the exceptions. Too often companies do too little too late. They know something is on the horizon but can't muster the

* Schmeck, H. M.,Jr., "Protein That Sets off Growth of a Human Organ Is Found," *The New York Times,* September 17, 1985, p. 1.

will to face it. And they have pride. Pride that borders on stubbornness. Consider the case of Transitron, a once hot Massachusetts maker of transistors. Most people thought Transitron, an early entrant into the transistor business, was an attacker. Instead, Transitron was a defender about to be caught in a transition.

As an important maker of germanium transistors, Transitron was in some strong markets, including military products, and growing fast. In the late 1950s, Transitron's sales increased sixfold, making it a hit on Wall Street, where its stock in one three-month period climbed from 36 to 60.

Then came silicon. In less than a decade it eliminated the need for germanium manufacturers. Transitron made a number of classic errors during this period of discontinuity. Its founders, the Bakalar brothers, acted as if they believed, on the basis of earlier growth, that success would be endless. Even as earnings began to erode they took on more debt so they could expand their germanium-based production. Silicon? The Bakalar brothers knew it was coming. They even set up a separate chemical company to sell silicon to Transitron. But their interest in silicon never amounted to much and that operation soon went into decline. In effect, the Bakalars made the same mistake American Viscose did when it tried to hedge on both rayon and polyester. Since both kept their bets small on the new technology, they never saw its true potential until it was too late.

Faced with a shrinking market for germanium, Transitron tried diversification, moving first into a variety of electronic products and then into, of all things, pup tents for the military, purchasing seven companies in that field. Presumably someone figured that if they sold the U.S. Army semiconductors they could also sell it canvas products. As its semiconductor sales declined, Transitron's profits vanished, from a high of $9 million in 1960 to an $18 million loss in 1969. At its peak Transitron's stock had sold at 56 times its reported earnings. But the Wall Street high-flyer soon fell from grace, crashing from 60 to 4 in barely two years.

For Transitron, the 70s and 80s were years of repeated failure and retrenchment. It tried making integrated circuits and microprocessors only to find the market always one step ahead. And that's typical. Maybe a company can produce one generation of a new product, but will it have the cash to produce a string of them? Not usually and not when it's bleeding in its mainline business.

Today Transitron is a $100-million company outside Boston. Not bad, but consider that when the semiconductor market was just beginning to grow, it was bigger than TI.

DEFENDER HUBRIS

Transitron exemplifies what I call defender hubris. It made a bad situation worse by committing a fistful of errors in a pattern that is common among defenders under attack from newer technologies.

The first common hubristic error is to believe that an evolutionary approach to technology will be good enough. Writing in the *Harvard Business Review* a couple of years ago, a senior executive at one of the country's largest companies contended that pushing the limits of technology is frequently unimportant because, he argued, customers will accept levels of performance well below those of the new technology. The failure rate for innovation is high, he went on, and our ignorance about a new technology is generally greater than we realize. What is more, he warned, the introduction of a new technology often requires lots of supporting investment and the culture required to effectively develop new technology encourages waste. His solution: avoid problems by adopting an evolutionary approach.

This executive, like many others, believes that statistically if you bet "no" most of the time you win. So they don't bet on discontinuities. Caveat innovator. Because the evolutionary approach is doomed to fail and cannot withstand the enormous rapid changes brought about by discontinuities.

TECHNOLOGICAL HEALTH

Hubristic error #2 is assuming you will have ample early warning about a coming discontinuity if you understand present technology, customer needs and competition. Since most companies don't study S-curves, technology limits and performance parameters, they don't understand where their own technology can go or what it will take financially to get it there. In a sense they confuse measures of economic health with measures of technological health. The simplest analogy I can give is when we are sick. If I don't have a thermometer, I can still tell I am sick by touching my forehead and feeling that it's burning. By then I probably should have been in bed hours if not days ago. The value of the thermometer is in being an accurate early-warning device, enabling me to take corrective action when things start to go wrong.

We don't yet have thermometers for measuring our technological health. Since most companies don't know how to measure their technological health, they measure their economic health. The trouble is that economic health is a result of many things that essentially are independent of the underlying technological health of the business. For example, a manufacturer can reduce unit costs by building a larger plant and spreading costs over a larger sales volume. An alternative is to wield the ax, cutting out the corporate staff and perhaps reducing R&D. For a while profits may improve. Wall Street may bid up your stock price. But these are transitory rewards. What isn't being addressed is the company's technological health. It is a bit like having a hot toddy when you feel a cold coming or drinking coffee when you are drowsy. You feel good for a while, but it is only symptomatic relief. What is worse, such actions delay a correct prescription. There is nothing wrong with symptomatic relief provided it is recognized for what it is. But when true medical relief is delayed too long, then the cold can turn into pneumonia. In business, as in health, getting better means fixing the fundamental problems, not just their symptoms.

KNOWING YOUR CUSTOMER

Hubristic error #3 occurs when a company is convinced it understands what its customers want. Talk to the owners of a failed business and invariably you will find them puzzled about how quickly they were forsaken by their customers. Do companies really understand what their customers want? Sometimes certainly, but often not accurately enough to know what those customers will do when approached by a discontinuity in products available to them. Customers are notorious about wrongly predicting what they will want. They did not want plain-paper copies—carbon paper was plenty good enough they said—until they could get plain-paper copies reliably and cheaply. They did not want "tummy" TVs until Sony opened up the market for them.

They would not trust paper diapers, until P&G made them strong, absorbent and cheap. So if the customer doesn't know what he wants until he gets his hands on it, it is hard to see how a company can put much faith in the thought that they will tell you that a change is coming.

Even if companies do know what the customer wants in general, it's not clear that these wants can be accurately and speedily converted into the specifications for products that economically meet the customers' needs. Each product has multiple performance parameters, some of which are in competition with one another. For example, the customer may want his copier to be speedier but not if that implies sacrificing reliability. Further there is often a complex relationship between what the customer wants and what the product designers can deliver. For example, for the copier, the designer may be able to provide the user with greater speed and equal reliability but only if the cost is increased substantially. So trade-offs are a way of life. Sometimes companies guess them correctly, sometimes they don't. Seeing changes in an already complex set of parameters that are traded off against one another is a difficult art.

Further, the discontinuity, when it comes, may not come in the main market sector, the sector where a company knows the most about its customers' needs. It will probably come in a niche. Nautilus equipment has many would-be challengers in the gym, but the big threat may come from a computer-based machine that doesn't use any cams at all. The computer allows a user to program a "curve" of resistance. But you won't find it in the health club. So far only 300 have been sold to the likes of NASA, Harvard Medical School, and the Denver Broncos.* Radial tires succeeded in

---

* Roberts, David, "Citizen Jones," *Ultrasport*, September/October 1985.

the replacement market first, not with the big automobile manufacturers. Most tire manufacturers assumed the customers would want to replace their original tires with ones just like them. They didn't. Detroit assumed that the customer would prefer a smoother ride to a longer-lasting tire. They didn't. Presently it appears that the non-caloric sweetener, Aspartame, is taking market share from hard-core sugar users as well as from the well-characterized diet market. That's an unexpected change—a discontinuity on the way. The point is even if a company knows its major market well, the discontinuity may not start there. It may well start in a niche where the company is not looking as hard as it can, and in retrospect as hard as it would have liked. So, knowledge of the customer is illusive at best and misleading at worst.

DEFINING THE MARKET

Hubristic error #4 is wrongly defining the market. Market definition is tricky in stable times, incredibly difficult in discontinuities. Instant cameras are an established product. Is video tape for recording and playing back television shows a potential competitor for Polaroid? We use Polaroids mostly to have a permanent record on photographic paper at home, whereas TV video tape cameras were initially used professionally or commercially.

But that's changing. In 1981, Sony introduced the electronic Mavica camera to record a still picture on a magnetic disk that is placed in your television set for viewing. Is that an instant picture? A year later Eastman Kodak came out with its own electronic camera. Now how do we define the market? Sometime soon camera markets may merge and Polaroid and Sony may be rivals.

When companies are going through transitions, it is much easier to misinterpret the market than to interpret it properly. Without the right data about the potential of rival technologies they don't have a chance. Companies overestimate their ability to see discontinuities coming because they believe they understand their competitors. But they don't and this is hubristic error #5. Often they don't know which competitors to watch. There will certainly be the behemoth whose every action is observed and reported, but often the most significant challenge will come from a smaller company. Armed with stealth, capable of higher technical productivity and unhindered by vested interest, it walks away with the trophy. Certainly this has been the case in electronics. But it also happened in medical instruments, personal computers and instant photography. It's happening now in pharmaceuticals, chemical synthesis, communications and information services. What is true for smaller, less-well-known companies is also true for foreign competitors. So the trouble with trying to anticipate discontinuities by watching competitors is that we usually watch the wrong ones.

Even if companies know who future competitors might be, they may not be able to ferret out their intentions or even the potential of their approach. To do that would require an understanding of their technology in some depth, and this may not be possible or likely. Even if the technology is known, it may not be easy to assess its limits or the likelihood of reaching them if it is based on an entirely different skill base and science. Can metal companies evaluate the potential of plastics or ceramics?

A company should, of course, always try to size up its competitors. But placing too much faith in the ability of corporate radar systems to pick up enemy planes may be too bold. Even if the corporate radars work, they may not work soon enough to allow sufficient time for a company to stymie an attack. Most companies think they can react fast enough, and this is yet another error of hubris (#6). Contrary to what most managers think, change occurs gradually only for a limited time period—until the attacker's product is economic in the customer's eye. At that point evolution ends and revolution begins. Many companies will disagree with this assertion, citing their ability to improve their own technology. True, to some extent. Defenders can fight back when challenged because in all likelihood they will be close to but not yet at the limits of their technology. The mistake comes in believing that this maneuvering will allow them to win the competition. It won't. It only buys time and maybe not enough to move onto a faster technological curve.

THE ATTACKER AND THE DEFENDER

A defender facing an attack can expect the onslaught to go through four distinct phases. First is the entry of a competitor with a new product into a market niche. New products are often like the foals of racehorses shortly after birth. Gangly and gawky and full of hidden potential that's going to take a lot of money to untap. Many more fail than succeed. But in some very specific segments of the market, perhaps the military or the Young East Side (Y.E.S.) market in Manhattan or Jaguar drivers, somebody is willing to pay the price for the product because it is new and different and in some ways better. And so one competitor, after several others fail, finally establishes a beachhead.

After niche entry the successful attacker spreads out and

begins to penetrate the market. At some point, all of a sudden, he hits it big. He takes a huge order. In the radial tire business it was when Michelin took the order for the Lincoln Continental. Suddenly, the French tiremaker was a force in the U.S. market, attracting attention and gaining market share. Before long it even started to dominate the market. In this third phase, the substitution is well along, actually moving toward completion. There are only a few niches of the market which haven't been penetrated. Then there is a fourth phase, when the remaining available niches are filled. These are different from the original high-priced niches the attacker went for. Often they are low-priced or segments with special needs. The attacker—soon to be defender—enters these through product variants such as sidewall radials, radials for older cars, radials for odd-sized wheel hubs, each attuned to an individual segment.

How does this four-phase cycle look to a defender? Initially the defender will see his own economic performance and sales continue to climb when the attacker is in the niche entry stage because the attacker is moving slowly. The defender is overconfident because he has seen several attackers fail. Even if a few succeed, their rate of penetration is often less than the rate of growth of the market. Even in a market that isn't growing, the attacker initially is taking only a small share. So the defender's economic performance is at best not affected materially, and at worst, in our sense of the term worst, is improving. It's impossible for a defender or investor to detect the potential of the attack by examining conventional economic indicators because the economic performance will appear to be strong or unchanged. It is the relative technological performance that is deteriorating. Thus, what defenders look at and treat as a proxy is the wrong proxy. They're collecting the wrong data.

As a result, the defender cannot see the attack coming, cannot count casualties or see shells popping. But out of his sight the attacker is gaining ground. He begins penetrating the market. He is no longer just in New York. He is also in Paris, Chicago, San Francisco, Los Angeles and London and he's becoming more noticeable. His rate of attack now approaches the rate of growth of the market. His costs begin approaching parity with the defender's. At that point, the defender's sales or market share may level off. Even then many businessmen still try to rationalize the situation, suggesting the problem lies not in new competition, but in the cyclical nature of their business.

It is around this point that the attacker comes into full view by winning the big order and the defender's sales begin first to sag and then collapse. Likewise his prices. The full attack is now on, and the economic indicators have caught up with the underlying technical reality.

Needless to say, things continue to deteriorate for the defender. Eventually he will leave the market and may even go bankrupt. In most cases, though, there is some residual market for the old technology. There will always be sailing ships. There are still some exotic vacuum tubes being made. The market served by the old technology will be small, have no growth, but may be highly profitable for the two or three producers left in. But as the industry as a whole experiences a shakeout, prices will collapse, and only a few firms will be economically strong enough to weather the storm.

It's possible to put this cycle of market penetration into a time frame. It usually takes between five and fifteen years for a new technology to supplant an old one. The first quarter of the transition period is taken up with the attacker moving into a niche. By the time the attacker has broadened his market, about half the substitution time is spent, and maybe two-thirds have passed when the defender really starts to feel economic pressure from the attacker's product. The full collapse follows shortly after. Thus, there are a few years in the latter part of the cycle in which the substitution is concentrated and quick. And by then it's too late to do anything.

This is exactly what happened in the tire cord industry. After a slow start against rayon, polyester had 30 percent of the market in 1970. It took about eight years to get to this point. It went to 70 percent in 1977, gaining about six market share points a year. Tire cords is about as slow an industry to change as you can imagine, tied as it is to the five-year model cycles of Detroit. And yet even here the leader lost half-a-market-share point a month during the middle of the transition. When radials finally caught on in a big way in the mid-1970s (Exhibit 20), bias-ply tires lost 50 market share points *in eighteen months*. That is almost three points each month. Almost three times as fast as the shift to polyester tire cords.

Radial tires have cords too. Originally, in the U.S. they were made of glass. Even before radials gained major share over bias ply tires, however, steel cords came in and dominated the market because of superior properties. The glass fiber manufacturers lost 40 share points in 1½ years, or 2¼ points of market share per month.

_____ THE SPEED OF TRANSITION

These rapid transitions have happened in many different industries, high and low tech. Once they established a foothold in part of the market, turbo fans took only six years to wipe out turbo jets. Second generation turbo fans took about five

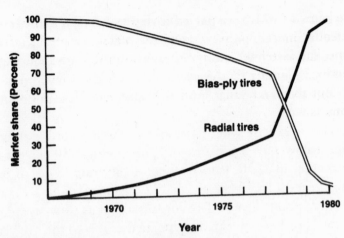

20 Tire Consumption in the United States.
Bias-ply manufacturers lost 50 percent of the tire market to radials *in 18 months.*

to six years to wipe out the earlier model. In each case, during the height of the transition, the defender was losing to the attacker at the rate of about one point of market share per month. As we noted earlier, electromechanical cash registers went from 90 percent of the market in 1972 to 10 percent in 1976, losing about 20 points of market share per year. Prices and profits collapsed sooner.

In electronics there have been many fast and traumatic transitions like the Transitron case. Germanium lost about 12 share points per month at the height of the transition to silicon. Integrated circuits went from 20 percent to 80 percent of the market in six years. Within integrated circuits we have seen rapid transitions from PMOS to NMOS to CMOS. In each case the attacking technology gained about 10 to 15 market share points per year after it caught on.

The speed of the transition is determined by the relative economics of the defending and attacking technologies and the rates at which the technologies move. The slimmer the defender margins, the greater the attacker rate of cost decline, the faster the economic transition will occur (See Appendix 3: Timing the Attack).

But there are other facts that also speed or slow transitions as well:

— The timing of the attack
— The relative value in use of the attacking and defending products
— The pricing strategy for the attacker and defender
— The extent of the need for the user to invest capital to take advantage of the new technology (the less additional capital required, the faster the transition)
— The opportunity for overall cost reduction in other areas of the user's business
— The general financial condition of the attacker, defender and customer
— The number of people involved in the decision-making process. (It is a lot easier to sell a product to a single company than to a joint venture marketing mainly to the U.S. government. That can take forever.)
— Finally, the spirit of innovation and entrepreneurship among current customers. (Are they psychologically inclined to be early users?)

The point is that there are often relatively long incubation periods in these transitions, whether one looks at sales or market share. Then suddenly the market collapses or explodes, depending on whether a company is the defender or attacker. Moreover, that period is probably getting shorter rather than longer for many products, since the rate of technological change is increasing and margins are even slimmer.

These events are not mysterious, nor should they be unexpected. The patterns are repeatable as Du Pont found out with nylon and NCR with electromechanical parts. Taking solace from the weaknesses of individual attackers, defenders avoid facing the reality that the attack is made up of not one but scores of attackers all hoping their time is the right time, that their efforts will be the ones that break the resolve and resources of the defender. Just as technological progress is inevitable so does it appear inevitable that the attackers will exploit the hubris of defenders until the defenders learn to see what the attacks truly represent—a challenge to the conventional order.

Joseph Priestly, the discoverer of oxygen and member of the Royal Society of England in the eighteenth century, used a scientific analogy to talk about the dangers he saw for the Royal Society. His words capture the flavor of present-day discontinuities. He said, "We are, as it were, laying gunpowder, grain by grain, under the old building of error and superstition, which a single spark may hereafter inflame, so as to produce an instantaneous explosion, in consequence of which that edifice, the erection of which has been the work of ages, may be overturned in a moment and so effectively as that the same foundation can never be built upon again."* More than 250 years ago, Priestly had an absolutely clear understanding of what is going on today. If we want to reverse this pattern, we need to start looking at the realities of discontinuities and their underlying economics.

* Center for History of Chemistry, University of Pennsylvania, CHOC News, Vol. 1, No. 2, Spring 1983, p. 1.

SEVEN

THE ATTACKER'S ADVANTAGE

The greatest inventive genius in recorded history was surely Leonardo da Vinci. There is a breathtaking idea—submarine or helicopter or automatic forge—on every single page of his notebooks. But not one of these could have been converted into an innovation with the technology and the materials of 1500. Indeed, for none of them would there have been any receptivity in the society and economy of the time.

—Peter Drucker, *Innovation and Entrepreneurship*, 1985

But the game isn't played in laboratories. It is in the marketplace. Isn't it here that big established defenders can crush upstarts, notwithstanding their technological advantages? After all, technological progress isn't the objective. The ultimate goal is to maximize return on a constrained resource, R&D funds, by improving products and processes and thereby capturing markets. The key questions are how do advances in technology translate into corporate success? And does the advantage that attackers have in developing new technology also give them an advantage in the market?

There isn't a great deal of evidence to answer the first question. When Bill Abernathy was at Harvard, he studied innovation in the automotive industry. He looked particularly at Ford and gathered data on the cost of a model change and the value to the company that resulted. He did not apparently try to take into account what profits would have been lost if the model change had not taken place, so his estimates are probably conservative. Nevertheless, it is the best data around. I used Abernathy's data to better understand the relationship between benefits (value) and cost of development at Ford. After some calculations I plotted his findings.

What they show is that around the turn of the century the

automobile industry was getting back $6 or $7 in return for every dollar invested in R&D. That ratio continued to increase in the early days of the industry up to a point in 1930, when the U.S. automobile industry was getting back $75 for every dollar invested in technology development. Then the returns peaked and started to come down. In 1945 the industry was getting back about $5 for every dollar it invested. Returns continued to slip from there. In 1965 the cost of developing a new model was about equal to the incremental benefit that the car makers received from developing it. They put a dollar in and they got a dollar back.

After 1965 they started getting 90 cents back. From roughly 1955 to 1975, the industry got back about what it put into development. It spent roughly $5.5 billion, which was economically unproductive—$5.5 billion out of $8 billion invested between 1910 and 1975. The first 30 percent of R&D spending accounted for virtually all the benefits the automobile industry has gotten back.

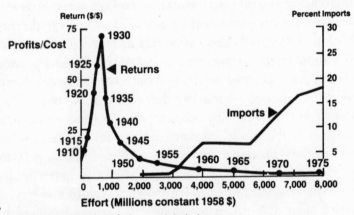

21   R&D Payoff in the U.S. Automobile Industry.
During the period of lowest R&D return, imports made their greatest entry into the U.S. market.

It was during the period of lowest R&D return that the Japanese made the greatest inroads into the U.S. market (Exhibit 21). Clearly the reasons for their success go far beyond technology. There are many things that enter into the relationship between technological progress and company profit.

_____ *A FORMULA FOR SUCCESS*

To understand it, it is necessary to think about the profit from investing in technology development in two parts. The first part is the amount of technical progress made for the investment. The second is the amount of profit made from that progress. We invest monies in R&D to make technical progress, and that technical progress gives us the potential, but only the potential, of making a profit.

To understand if profits will be made, we must turn the spotlight on the second part of return, the amount of money made from a given technical advance. Call this ratio R&D Yield. The overall return on R&D, the amount of money we make from a given technical investment, is equal to the amount of technical progress we make from our investment (R&D productivity) multiplied by the amount of money we make from advancing the technology (the yield).

Expressed mathematically:

R&D Return = R&D Productivity × R&D Yield

In Chapter 4, we saw how attackers with new technology typically have a 5-to-1 advantage in productivity over incumbents with old technology. And in some cases, for example in the electronics industry, it was 20- or 30-to-1. But what about the yield side of the equation? Can we understand that side as well, and what will it tell us about R&D investment strategies?

We can understand by applying basic microeconomics. Yield is a function of the value of a particular product to a customer and the degree to which its manufacturer can protect that value from competitive duplication. Just as with productivity, this can be quantified.

It's easiest to think about the value as lower cost. If advanced technology allows a company to make the product for a lower cost, then it makes more money if it charges the same price. On the other hand, maybe it can enhance the value by adding product attributes that the customer will pay more for.

Consumers in the marketplace don't care whether the company spends a dollar or a billion dollars to develop a new product. They care only about product availability, its technical advantage over alternatives, and how many companies can supply the product.

So it makes sense to separate productivity and yield. We can then think about the R&D return in a structured way. We can also get information for each part from different people with different skills. We know what information each must supply. R&D must supply knowledge of the basic technical alternatives, what their limits are, where each competitor is on its S-curve, and what can be done to improve the slope. The business planning/marketing community must bring knowledge of customer values, future supply and demand patterns, plus an idea of likely competitive strategies and prices and how these might change the market structure.

Discussion between the marketers and the engineers often become bickering matches. Marketers complain about technical successes that were market failures, and engineers complain that marketing doesn't know how to sell or what customers want. Partitioning the equation for R&D return helps technical people develop a clean sense of what they have to contribute. We expect marketing executives to know the size of the market. In a similar way we should expect technical people to know the limits of their company's technologies.

When the CEO of one of our clients (a $5 billion manufacturer of electronic gear) first saw the partitioning he said, "Now that shows us where we've gone off course. We're getting less and less out of our technical efforts (the productivity is low), but we're still making a lot of money (the yield is high). Everything looks okay, but it's clearly not. We're probably close to the limits of our present approaches. Productivity may go to zero, so we have to think about changing our approach." Given the way he phrased the proposition, it was clear who had to do what. Specifically, the technical community had to see how much further present productivity could be improved without switching to a new technology. That effort has now been completed and it resulted in a major reallocation of technical investments from the old technology to new ones. And that is starting to have a major impact on the company's competitive position.

Top management must ask some tough questions. Is there a common understanding of the technical performance parameters? Do changes in technical performance factors come in eras? If so, what have they been and how frequently have they changed? What are the forces leading to the change? How might they change in the future? What might future performance parameters be? Is R&D working on improving today's performance parameters, or yesterday's or tomorrow's? Is marketing conceptualizing the marketing promotions, positioning and advertising campaigns in terms of today's or future performance parameters? Armed with the answers, management can think about the return on technical investment. Whether they do it analytically or intuitively isn't important. The framework helps theoretically and practically because it substantially improves communications.

Breaking up the equation can also help us see what some companies do wrong and others do right. We want the return, that is profit over R&D investment, to be positive. This can happen only if both the R&D productivity and the yield ratios are positive. In real life, however, either can be zero or negative. Zero productivity means a company is not making any technical progress from its investments. Negative productivity occurs when it tries to improve one attribute of a product, say its reliability, but does so by sacrificing the performance of another attribute, say speed, with the net result that the overall product is less appealing to the customer. Yield is zero when no profit results from the technical progress, and negative when the more advanced product earns less money than its predecessor, which might happen if the market suddenly became oversupplied with the more advanced product. This happens frequently in the electronics business. Some of the newer products make less money than the older standby products.

Since either productivity or yield can be negative their product—R&D return—can be negative. We can have negative returns when R&D productivity is high but yield is negative. This was the case when detergent makers spent a great deal of money putting in expensive optical brighteners that failed to produce a meaningful benefit for the consumer. Clothes were "brighter," technically, but not to the naked eye. Productivity was high, but return was decreased, because market share stayed the same and cost went up.

Of course, another situation to avoid is where the yield is high but the R&D productivity is zero because the technology is close to its limits. That, as we have seen, often happens. Some steelmakers directly reduce iron ore and other products (like carbon) to steel rather than going through the indirect process of first producing less pure, or pig iron, steel. The direct process is economic only with an abundant supply of low-cost natural gas, as in Mexico or Indonesia. One well-respected Mexican steel company used it to capture a large market share until challenged by a newer process developed by a German firm. The Mexican company tried to improve its process but with scant success. It was near its limit. Meanwhile the Germans were selling steel at prices that the Mexicans suspected were below cost. They said, "The Germans are destroying the market. They never care about profit. We must slash our own prices to compete with them." After a long and costly period, one of the Mexican engineers realized they were wrong. The German process had different underlying thermodynamics. They could be much more efficient. In fact, they were making a profit, despite their low prices.

The failure of the Mexican engineers to study the limits of their technology cost them several years of wasted development effort as well as lost market position. The minute the Mexican company saw the problem, they realized that another process that they had under development, but hadn't put much effort into, had much higher limits and higher potential. They immediately switched their efforts to the new process and stopped development on the old. Within months they had a process that was superior to the German steelmakers' and were on the way to one of Mexico's biggest commercial successes. The point is that investing in technologies that are virtually at the limits makes as little sense as investing in technologies that have lots of technical potential, but have little value to the customer. What we want are situations where both the R&D productivity and yield are positive.

THE DEFENDER'S DILEMMA

There are other factors affecting yield including a company's cost competitiveness and the collective strategies of all the companies in the industry. Consider, for example, a reasonably competitive industry where supply and demand are nicely in balance. Everybody is making money. Under these conditions one company decides to invest further in the business. Trouble is, everyone else is thinking the same way. Everybody puts up a plant. The industry now becomes oversupplied. The market needed only one more plant, not six. The power shifts to the buyers. Prices collapse. And no one gets the expected return from R&D investments.

This is what happens all too often in the chemical industry. When phthalic anhydride changed its technology from the naphthalene to the orthoxylene process, demand was at a peak, output at capacity and prices skyrocketed. Everyone wanted to build a new plant, but the backlog in construction orders was huge. Nevertheless, in a four-year period, no fewer than eight plants were built. Supply rose 80 percent, but demand for phthalic anhydride increased only 58 percent. Customers could see what was happening and they started negotiating for lower prices. In the next three years prices halved even though overall demand grew by 15 percent a year. The return on all that investment in the orthoxylene process? Minimal, at best.

Assume we are in an industry that has six plants with the supply and demand curves looking as they do in Exhibit 22 and that we own plants one and four. We have the lowest-cost plant and own a medium-cost plant, number four. Basic economics tells us that we find the marginal supplier where the supply and demand lines meet and that the production costs of the marginal supplier are equal to the price of the product in the market. All those competitors with costs less than the marginal competitor's will make money while the rest lose. Actually they don't lose money; they stop producing at least until demand picks up.

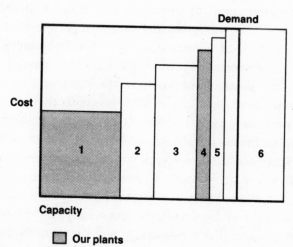

22 Supply and Demand Curves.
New capacity gives buyers the power to negotiate lower prices.

To see the connection to yield, assume we have invented a new low-cost process for making a standard product. When we build our new plant, we add capacity to the industry. If demand hasn't grown (and for the sake of this simple example, let's assume that is the case) this new capacity isn't needed. Or to be more specific, since our process costs less, the capacity of the marginal producer is no longer needed. We create a new marginal producer, with lower costs than the old one. Since price is equal to the cost of the marginal producer, and our entry with a new process caused a shift to a new lower-cost marginal producer, prices in the industry will fall.

This is common experience. Prices will fall by an amount which is proportional to the amount of capacity we've added. The more we add the more prices will fall (the proportional constant being determined by the cost structure of the industry). In real life it's even worse, because prices don't fall when the plant starts to operate, but when the plant is announced, or even when rumors of an announcement begin to circulate. If you doubt this, think what happened to personal computer prices when IBM announced its entry into an already crowded marketplace or what has happened to electronics prices as the Japanese have entered the business.

The price decline has to do only with the amount of capacity we add. Except in the extreme, it has nothing to do with the technical advantage we have. In our example our new process not only knocks down our competitors' price, it also affects the profitability of our existing plants. Indeed if one of our plants was the marginal producer we may cause that plant to shut down entirely because of our own actions.

Think about it this way. How will the proposition look to someone who is not a current producer and has no investment to protect? It might look pretty good. Their entry probably will make market prices fall, but their newer process will still yield a high enough return to justify the investment. We are not asking the right question if we want to know only if our own investment justifies obsoleting our existing plant. That contains an implicit assumption that it is up to us to decide if a new plant enters the market or not and if prices will hold or not. If there is an incentive for someone to get into the marketplace, then someone, someplace, someday, will do it. At that point prices will decline. If we sit on the sidelines we will lose the profit we sought to protect. The only case where this wouldn't be true is if the costs of learning the business more than offset the potential profit. That happens typically in technically complex businesses such as turbines, CAT scanners and jet engines. But try recommending additional capacity at the next budget meeting when the old is sufficient. Chances are you have just lost your next promotion.

If a company expands in small increments in step with the demand expansion of the industry, then profits will climb as technical performance goes up. But if big expansions throw an industry into overcapacity, there will be sudden and precipitous drops in profit regardless of technological gains. Profit as a function of technical performance is not some unique relationship we can determine in advance like R&D productivity. It's a function of the microeconomics of the industry and the collective strategies of everybody in the industry. There is nothing inherent about the relationship between technical progress and profits, which means we must analyze the problem from a fresh point of view each time. But this much is clear, the yield will look greater to a company without current investments in the business than it will to one that is worried about protecting those investments. With yield the attacker has a perceived advantage.

Let's expand the scope of this discussion to enhance our understanding of the economics of substitution and how they favor attackers. Assume that the constrained resource is capital. Now the goal is not to maximize the return on R&D investment, but to maximize the return on capital. To look at the problem this way we have to discuss one more ratio—the R&D multiplier. This is the ratio of R&D investment required to develop a product to the capital necessary to make and sell it. For example, an R&D program costing $1 million may require, if successful, the company to spend $10 million on a new plant. In this case the R&D multiplier is $1 million/$10 million or 0.1. If the $1 million of R&D required $100 million of capital the R&D multiplier would be only .01. We call this ratio the "multiplier" because if you multiply it by the R&D return we get the return on capital for the business. Again we can write an equation.

R&D Return × R&D Multiplier = Return on Capital

Given this scheme, the smaller the multiplier, the lower the return on capital. To put it simply, a company would rather spend a dollar of capital and not $10 or $100 for each dollar of R&D. If new technologies require big investments in new plants, they are not as attractive.

The range of these R&D multipliers is startling, varying by a factor of 100. Exhibit 23 shows R&D as a percentage of capital expenditures, which is another way of thinking about the relationship for selected industries. Aerospace companies, for example, spend over five times as much on R&D as they do on plant and equipment. In other words, they can advance technology without concomitant heavy capital commitments. On the other hand, iron and steel com-

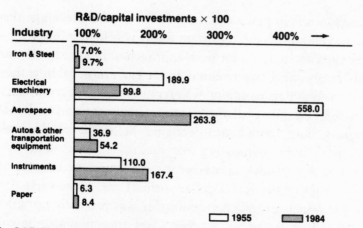

23 R&D Expenditures as a Percent of Capital Investments.
Aerospace companies spend over five times as much on R&D as on plant
and equipment. Iron and steel companies, on the other hand, spend *less
than one-tenth.*

Source: McGraw-Hill Economics, May 1984.

panies do just the opposite: $14 of capital for every dollar of
R&D. What this means is that some industries need to get
much higher returns on R&D to get an acceptable return on
capital. Paradoxically, these industries often believe implicitly that they are close to their limits and have low R&D
productivity. They are thus technologically conservative,
spending little on R&D. They become vulnerable to unexpected challenges from outside their industry, as steelmakers
have become vulnerable to plastics manufacturers.

Let's imagine we were investing during the middle of the
discontinuity from vacuum tubes to transistors. In this case,
there was a lot of technological potential left in both technologies. While the productivity was higher for the transistors, the yield was probably lower because there was a lower
sales base. The productivity in vacuum tubes was lower but
still substantial, and the tubes had a high yield because pro-

duction was in place and manufacturers could leverage their sales volume. So the return on R&D was high in both cases, perhaps about equal for both approaches. But the R&D multiplier probably wasn't equal. In the early days of the transistor, a great deal of capital was required for each dollar spent on R&D because all the process equipment had to be designed, built, tested and debugged before it worked well. Vacuum tube equipment on the other hand had been undergoing steady improvement for a long time. Less capital was required per R&D dollar spent. Thus, in the early days of the transistor its R&D multiplier was probably far worse than it was for vacuum tubes. Over time though, the cost effectiveness of vacuum tube equipment stayed reasonably stable whereas that of transistors strengthened a good deal. Ultimately, less capital was needed in the transistor business than in vacuum tubes to follow up the R&D. Return on capital went up and the discontinuity was in full swing. The number of transistor manufacturers exploded. Today chipmakers are building high-volume plants that cost over $100 million. Even though they are spending considerable sums on R&D, the multiplier may be much smaller. But process technology is changing so fast that companies have continued to build new facilities at a rapid pace.

In certain industries, notably chemicals, petroleum and pharmaceuticals, the R&D multipliers for the new and current technologies are similar. It's possible to reach sound conclusions about the best investment by considering the R&D returns alone. That is not to suggest that companies substitute return on R&D investment for return on capital as the primary measure of corporate performance. The ultimate long-term constraint in most companies is capital and high returns on capital are the objective, not high returns on

R&D alone. But in many cases the two points of view are consistent.

Thus, it is often the case that attackers have both a *real* productivity advantage and a *perceived* yield advantage. When multiplied together, these two components of our equation of return present a formidable challenge to the management of incumbents—one that they cannot hesitate to respond to because time is often short.

Imagine a situation where a new product enters the market against an existing product and that over time neither the cash (variable) costs nor the total costs of the current product change. Further, assume that the new product is more expensive (in both cash costs and total costs) than the present product, but these costs are falling rapidly because of advancing technologies (see Appendix 3). At some point, the total costs for the new technology will be equal to, and then less than, the total costs for the current technology. If the cost reduction for the new product continues, as it does during many discontinuities, the total costs of the new technology will be equal to, and then less than, the cash costs for the current technology. This is a grave situation because it means that the new competitor can build a plant and recover his total costs, at prices that won't permit the current competitor to even cover his cash costs. If he stays in business, he'll go bankrupt. More likely he will drop out shortly after he realizes his position. The discontinuity is now over, both economically and technically.

What is perhaps surprising is how quickly this can happen. It's not atypical for the cash costs of a product to be around 60 percent of the total costs. This would be viewed as a good business. It provides a 40 percent margin for all other costs. Assume that the total costs of the new technology are coming down at 20 percent a year, which is typical, perhaps a bit conservative for electronics products. How long will it be from the time when the discontinuity begins (total costs equal for both products) to when it ends (total costs for the new product equal to the cash costs for the current product)? Two years. At the end of the first year, the attacker technology can take the market price down by 20 percent while covering his total costs. This cuts the previously high margins of the current producer in half. He may feel the pressure and try to respond. But by the end of the second year, the attacker has taken the price down by another 20 percent and brought the ax down on the defender's current product. These economics, then, explain the rapid changes in market share discussed at the end of Chapter 6. The cycle of incubation and rapid explosion or collapse of share (depending on whether a company is the attacker or defender) is not a fortuitous event. It is a consequence of the underlying economics. As such it can be understood, analyzed and anticipated.

The underlying economics are what drive the rapid substitution of one technology for another. In PA the difference between total cost and cash cost was 30 percent. The costs of the new technology relative to the old were coming down at 7 percent per year. Thus, the transition took about four years. During this period naphthalene's share went from 80 percent of the market to about 30 percent (Exhibit 24).

_____ HOME ECONOMICS

There is perhaps one more economic advantage attackers have—call it home economics. Attackers are often little companies or small divisions of big companies, whose whole life

PHTHALIC ANHYDRIDE FEEDSTOCK
Naphthalene vs. o-Xylene

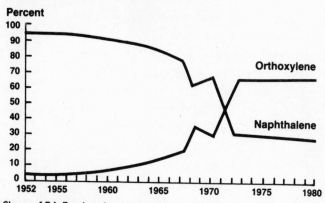

24 Share of PA Feedstock.
Propelled by the underlying economics, orthoxylene's share of the market for phthalic anhydride increased from 30 percent to 80 percent in about four years.

and fortune—house, spouse, kids, everything—are dependent on making the new product competitive. Most defenders find their cash flow comforting. They just don't have the desperation or feel the white heat of competition as do the attackers. Combine their complacency with disadvantages in productivity and yield, and it adds up to an overwhelming advantage for attackers.

Fortunately, there are some inherent attacker weaknesses. But only a few. They are not enough unless companies guard against the errors of defender hubris, understand the limits of their technologies, and are resolute in their determination to simultaneously attack and defend themselves.

EIGHT

COUNTERATTACK:
THE BEST DEFENSE

In its most famous business, run-of-the-mill business computer systems, IBM will soon cease to be a significant force.... Hundreds of new companies are bursting into existence.

—Adam Osborne, founder, Osborne Computers

On September 13, 1983, Osborne Computers filed for bankruptcy. Customers had stopped buying the current generation of Osborne computers, waiting instead for the new, improved products Osborne had announced but couldn't deliver on time. So attackers do not always reign victorious. The fact is attackers can and do lose. They lose frequently even when defenders have psychological and economic disadvantages. Or to express that more positively, smart defenders find ways of thwarting their attackers, at least for a while.

_____ DOGFIGHT IN THE SKIES

The development of the commercial jet plane is a case in point. There were several efforts after World War II to develop a commercial jet. The British had one of the first models, the Comet, that was based on a standard wing design with jet engines affixed. The catch was that the design wasn't stable at high speeds. The plane shimmied and shook. The project was scrapped. The leading U.S. plane builders, McDonnell Company, Douglas Aircraft and Lockheed, observing the British failure, put their plans for a commercial jet plane on the back burner and stayed with propeller-driven planes.

Boeing, then a weakling in the industry, had a better idea. Its engineers gambled that a swept-wing jet plane would be more stable. It was. Boeing shortly introduced the 707 and a once frail company became the industry leader. From that point on its competitors were playing catch-up. From the successful introduction of the 707, Boeing had enough cash to immediately start the design and prototyping of their next aircraft, the 727. With relatively limited resources they brought out the 727 just as McDonnell Douglas (they had merged) came out with the DC-8. The 727 was a huge success—the only plane in the last 25 years to really make money.

By the mid-1970s, however, Boeing had become a defender under attack by Airbus Industrie, the European-owned maker of commercial jet liners. As an attacker, Airbus Industrie scored a significant and early victory manufacturing the only two-engine widebody in the skies from 1974 until the Boeing 767 was put in service eight years later. The new design gave the Airbus much better fuel economy. In three years Airbus Industrie's market share soared from barely 3 percent to more than 38 percent, and in 1981 Airbus Industrie actually moved ahead of Boeing in new orders written. Many factors determine the choices airlines make when they select new planes, but in recent times few have ranked higher than fuel efficiency and financing. Airbus Industrie scored in both areas, offering the airlines great savings in both fuel and money costs.

Boeing, now the defender, has struck back, offering improved designs and extensive product support in the form of free spare parts and training, plus the promise of an earlier delivery date than Airbus for the next generation of mid-size commercial jet liners.

Boeing has played a canny game. While it appeared slow and hesitant, it in fact was handling its engineering and testing programs in a manner that would allow it to at least catch up and possibly bound ahead of Airbus Industrie. Boeing spent about 2½ times as much on the preliminary designs as did Airbus. It was worthwhile. Boeing began major assembly several months after Airbus, but went to first flight several months earlier. It spent more on the cheaper design phases, but overall spent less by keeping its costs low on the expensive construction phases. The payoff: United Airlines, TWA and Delta are just three of the airlines which have given Boeing orders that Airbus once expected to get.

The outcome of this battle in the skies now hinges as much as anything on yet another dogfight—the contest between engine makers such as Pratt & Whitney and General Electric.

Airbus is spending $4 billion on a new 164-seat plane, the A320. The first models will be powered by a fan-jet being made jointly by GE and France's SNECMA. Later models likely will carry engines designed by a Pratt & Whitney–led consortium. Deliveries start in 1987. That is the year GE and Boeing expect to make a decision on a revolutionary new engine. GE's new propfan engine promises to deliver the speed of today's jet engines but without burning nearly as much fuel. In fact, GE claims its new engine will have more than twice the fuel efficiency of today's engines and probably 25 percent more than any other engine now being worked on. The savings to the airlines could be enormous. Maybe as much as $60 million a year for each of the larger airlines if the price of oil stays at $25 per barrel. Trouble is, the new GE engine is still unproven and is not expected to go into operation before 1991, several years after the next Airbus climbs into the skies powered by the Pratt & Whitney engine that promises to be some 14 percent more efficient than to-day's best. Will the airlines wait around for the GE–powered Boeing or will they opt for the potentially somewhat less efficient P&W–Airbus plane? For Airbus the answer to that question could decide its fate. In effect, Airbus has bet the company on its next product. If Boeing, the defender, can outmaneuver it, Airbus could find itself with serious financial problems. Whatever the outcome, the moves and counter-moves that unfold are likely to be typical of what happens when attackers and defenders lock horns.

_____ *NOT EVERY ATTACKER WINS*

Earlier I made the claim that maybe as many as seven out of ten defenses were unsuccessful. But that does not mean only three out of ten attackers fail. Far from it. To begin with, a single attack rarely brings outright victory. So attackers must make repeated forays to wipe out the defender. It is rare that a single company has exclusive rights to a new technology. More often a new technology means an array of attackers. In the early 1900s, there were no less than 250 automobile manufacturers. At the peak of the germanium business there were more than 50 germanium producers all trying to cap-ture the vacuum tube market. Today there are more than 200 biotechnology companies trying to wipe out the leading pharmaceutical companies. And who knows how many mi-croprocessor companies are striving to defeat Intel or TI.

    Attackers blunder as often as defenders, maybe even

more often. By nature they are opportunistic, wanting to run before they can walk, thirsting to make the big kill. The rationales, conceptualizations and strategic thoughts emerge only shortly before the annual report goes to press. Their salesmen don't get useful information from their customers and channels. Their marketing and R&D and manufacturing departments don't communicate. Plant equipment is not available for test runs of new products. Their competitors aren't always identified, much less analyzed.

Witness the fate of Gavilan Computer. In 1982, Manny Fernandez, former president of microchip-maker Zilog, raised $31 million to start Gavilan as a producer of portable computers. After little more than a year, Fernandez was showing his first products at the trade shows. Buyers were enthusiastic. But then the problems began. Other companies including Compaq, Hewlett-Packard and GRID sensed a potentially huge market—over $4 billion by some estimates— and saw scant competition. They began developing their own portable computers. To get into the market quickly, Fernandez had arranged with Hitachi to supply disk drives for his machine's 3-inch magnetic disks and to have software compatibility with Apple's Lisa. Hitachi couldn't deliver on time and Lisa flopped. Worse, the rest of the computer industry was now using 3½-inch disks and had compatibility with IBM. Suddenly Fernandez's once-hot portable computer was seen to be operating outside the industry's new standards. Fernandez tried making changes, even raised more capital, but couldn't stop the bleeding. In October 1984, Gavilan declared bankruptcy.

Hundreds of other would-be predators have lost control of the attack process and failed. There are any number of reasons why this happens. Companies fail when their attack depends on expanding their area of expertise. The crackerjack company noted for its contract research work may quickly go broke when it tries to set up its own manufacturing operations. Poor engineering design regularly takes its toll. Machines have too many parts, making them hard to assemble and service. Too much inventory. Not enough automation. Poor finance. Too much debt. Too little equity. The wrong choice of distribution channels. For example, the early makers of digital watches tried to enter the market through the traditional jewelry channels and failed. The jewelers did not want to jeopardize existing business by accepting lower-priced competing products. A costly error. Eventually a salesman took his digital watch to alternative channels, first to drugstores and then to hardware stores. Posh jewelers had to scramble to get back into their own business. But digital manufacturers lost valuable time that some defenders, such as Texas Instruments, were able to use to get on board the new technology.

So even though attackers, as a group, more frequently win than lose, any individual attacker will find it tough to win. The odds against success are stacked high, no matter which camp you are in.

_____ THE DEFENDER'S DEFENSE

Looked at from the defender's point of view, the death rate of entrepreneurs is very high. Two out of three in the electronics industry for example. But by focusing on the odds that an individual attacker will get knocked off, the defender may miss the central point that the overall attack succeeded. The order of things changed. Typically the defender will be participating in a business that has settled down to a limited oligopoly with maybe ten competitors, of which three or four dominate the industry. There will be fifty or a hundred or as in the case of automobiles, biotech and the microcomputers, two hundred attackers. If the success rate for attackers is one out of three, or even one out of ten, there will be a substantial number of surviving attackers. They are the ones that will do damage to the defenders. The attackers, as a group, will dominate with a few of the defenders. Most of the defenders will be gone.

But as the Boeing Airbus battle illustrates, defenders can strike back. There are generally three different ploys used by defenders to thwart attackers—leapfrogging to a new S-curve, adding "sails" to the current one, and deep pockets. Boeing, for instance, is a company with both deep pockets and the skill to leapfrog. So was General Electric when for a while it faced an unexpected challenge in medical technology from Britain's EMI Ltd.

For many years EMI had been a modestly successful London-based record company catering to the tastes and whims of British, and to a lesser extent, American teenagers. EMI broke into the big time when its U.S. subsidiary began marketing Beatles records. Thanks in part to the Beatles and their prolific record sales, EMI in the 1960s generated sufficient cash flow to warrant a diversification program that moved it into electronics and the manufacture of components.

By chance an EMI engineer happened to live next to a brain surgeon, who often complained his work was hampered because X rays couldn't "see" brain tumors. The engineer, Godfrey Hounsfield, was intrigued—so intrigued that he eventually devised a way to combine X rays and mathematical analysis to locate tumors. His idea became known as Computer Assisted Tomographic Scanning, or CAT Scan-

F:I-7 ·

ning, for which Hounsfield shared a Nobel Prize and gained a knighthood.

For EMI, CAT scanning meant a whole new business and the opportunity to build medical devices for hospitals throughout Europe. Almost overnight, EMI transformed itself into a high-tech sensation. It had outmaneuvered the traditional suppliers of X-ray equipment to radiologists, doctors and hospitals—companies such as General Electric, Siemens and Philips. EMI was the toast of investors on both sides of the Atlantic. Its prospects appeared limitless.

But in the fast-moving, high-priced world of medical technology, EMI's success was short-lived. As an attacker, EMI was ill-prepared to combat the leapfrogging tactics devised by its opponents. EMI's CAT scanner was a remarkable advance over the existing X-ray equipment, yet it was only in the early stages of its own development—at the beginning of its S-curve. It was relatively easy to improve technically, provided a company had the money for the necessary development and process work. Unfortunately, EMI was technologically rich but cash poor. It could not afford to keep pace with the innovations that the better bankrolled GE, Siemens and Philips could afford. Before long EMI was trying to spend more than its total earnings on developing the next generation of CAT scanners. As its orders dwindled and profits disappeared, EMI could only watch as its competitors leapt ahead with improved CAT scanners.

In 1979, the year Hounsfield received his Nobel, EMI was bought by Thorn Electrical Industries, which quickly sold the CAT scanning business outside North America to General Electric for $32 million and the North American business to OmniMedical for a modest $5 million. EMI's attack had failed. Or looked at from another perspective, the counterattack had succeeded.

In most cases defenders can leapfrog when the attacker's product or process is far from its technical limits. In other words, where there is high technical potential and the costs of bounding up the S-curve are not high. Electronics, biotechnology, materials science, medical care and artificial intelligence all fall in that category today. Conversely, leapfrogging does not work very well where there is limited technical potential, where the cost of closing the gap is substantial and where the incremental gain over current competition is small. Predictably, in periods of leapfrogging, prices come under pressure as the defender's volume falls or as an increasing number of companies swarm into the market.

Leapfrogging works best when the cost of imitation is much less than the cost of invention. In military procurement there is some evidence that the cost of developing software, say a flight collision avoidance system, goes down 90 percent between the first and second application and then declines a further 90 percent between the second and fourth applications. By waiting until the second generation an imitator can save lots of money. These economics are the ones that drive attackers to raid defenders. Fairchild lured Lester Hogan from Motorola, Motorola took Al Stein from Texas Instruments, Genentech hired Kirk Raab from Abbott Laboratories. It can be worth a great deal to raid.

Just the opposite is true when the cost of invention is less than the cost of imitation, as it is when the imitation is prevented by patents or trade secrets or by high entry costs. In these cases it pays to do the research necessary to continually position the company to stay ahead.

Defenders can buy time to research and begin a transition by figuratively adding sails to their products just as the nineteenth-century shipbuilders literally did. It is a form of counterattack that can delay (but not ultimately prevent) the attackers' advance. Both General Motors and Ford have thwarted attempts by attackers to introduce new types of automobile engines—for example, gas turbines—by improving the classic (Otto cycle) approach to engine design. Similarly, the metals manufacturers have limited the use of plastics or ceramics in cars and other products by introducing high-strength, low-alloy steel in metals. A "fluidized bed" process has lowered the cost of burning coal cleanly and may keep coal-burning utilities from being replaced by nuclear-powered ones for some time.

Attackers get "sailing shipped" when they make an unrealistic assessment of defenders' ability to improve the cost effectiveness of their products. It is hard for an attacker to understand a defender's technological limits because his own skills are usually quite different. His strength in attack becomes a weakness against counterattack if it blinds him to what a defender might do to strike back. If he is to be successful, it must be difficult for the defender to replicate the performance of his product or service for a similar cost. Competitive analysis is key to making the right decision on market entry. Often, this analysis will require the gathering of new technical data that can be done only by laboratory work. This work may require technical skills not available in-house. But places like the Battelle Memorial Institute or Midwest Research Institute or Research Triangle Institute in Raleigh-Durham, North Carolina, are set up to supply them.

Many attackers make a fatal mistake early in their careers. Worried about the defender's ability to keep adding sails, they decide to bet the company. They hold off coming to market until they have designed what they believe is the ultimate product. This works if there is not much potential to advance or enhance the defending technology. But if a defender can advance the technology incrementally, inch by inch, protecting his market, his customers and his image along the way, the attack may fail. By the time the attacker comes out with his "killer" product, it is barely a threat.

_____ HYBRIDS

An important variant of the sailing ship counterattack is the hybrid product (or process). Successful defenders often will take elements of the newer technology and add them to the old. Literally add a steam engine to a sailing ship (Exhibit 25). This allowed the ship to sail quickly when there was wind, but not to be powerless should the "doldrums" appear. For a period the hybrid was more cost effective than either the "purebred" sailing ships or steam ships. Hybrids operated for almost 100 years—a long discontinuity.

We saw hybrid products, too, in automobiles. Even though they were first called "horseless carriages," there was a substantial period of time when a "back-up" horse *had* to accompany the car. In tires, we did not go from bias-four-ply tires to radials. We went to bias-belted radials and then to radials. Electronics has had many hybrids such as radios and TVs that contained both tubes and transistors. Aspartame, or NutraSweet to most of us, was first marketed as a hybrid with saccharine.

We often don't know what to do with a new technology or materials. The first cast-iron bridge, which still stands in

25 The Rattler, 1845.
A hybrid. Is it a steaming sailship or a sailing steamship?
Source: Angelucci, Enzo, and Cucari, Attilio, *Ships,* McGraw-Hill, 1975.

Great Britain, is a beautiful structure, but it is a cast-iron bridge in material only. It is built exactly like a wooden bridge, right down to the pegs used to hold it together. Why? Because no one knew then how to plan and build a cast-iron bridge. It was more than a bridge across a river; it bridged wooden technology with iron.

Hybrid products are usually most successful in specialty markets. The sail/steamship hybrid was valuable for long hauls of costly cargoes through areas where the winds could die, such as a trans-Atlantic voyage, but less valuable hauling a low-value cargo around the Caribbean, where the trade winds blow and where the value of the fuel would be higher than the product carried.

Hybrid products represent a compromise between a premature new technology entrant and a maturing current technology in need of aid. They can be a way to quickly and cheaply find out about product problems through the market rather than through research. And a way to temporarily bolster sales and market position. But many companies do not try to design hybrid products. Instead they either try the pure sailing ship approach or try a direct counterattack with the new technology. The "all solid state" TV, the radial tire, the anticancer drug, the full-blown steamship. But the hybrid approach often can be done quickly and cheaply.

In response to Michelin, Goodrich tried to sell Detroit their brand of radials, but Detroit was not buying. Car makers had not had time to redesign the shock absorption system for the new rougher-riding radials. Goodyear succeeded by introducing its bias-belted radial, which was actually a hybrid product—a cross between the conventional bias-ply tire and the newer radial. The big advantage of the bias-belted radial was that it was acceptable to Detroit and could be made on conventional bias-ply tire-making equipment changed only slightly, and at low cost, to accommodate the new designs. All this served Goodyear very well in the late 1960s and early 1970s and helped stave off the penetration of Michelin.

But as we know, however, radials won out. Goodyear then had to go all the way to radials, which cost them dearly. Their hybrid product, as a stopgap, worked well. It was not the whole answer, however. In this sense hybrids can be dangerous if management lets them drain resources from what is crucial in the long run—learning how to attack with a new technology.

There are several good examples of defenders who know how to counterattack. Consider what has happened since IBM began counterattacking competitors in both personal computers and in disk memory devices. One victim, Storage Technology, filed for bankruptcy in October 1984, less than two years after its revenues topped out at $1 billion. Wang hasn't gone bankrupt, but its earnings have declined since it began to go head-to-head with IBM in office systems, its newly appointed president has gone off to do other things, and retired chairman An Wang is back running the company.

IBM does its homework well and knows how to retaliate in ways that test its competitors' weakest points. It knows how to use its resources to improve products and/or reduce prices. It is also skillful at other countering tactics such as using announcements of coming technological change to encourage customers to delay their purchases of competing brands and models until IBM's are in the marketplace.

Until its head-to-head fight with IBM, Wang had successfully defended itself several times in word processors and small computers. For example, in 1975, DEC introduced its new small computer system, the "System 310," and IBM, struggling to reestablish its position in the market, came in with its System 32. Wang was threatened. Immediately, however, it came back into the market with its Wang Computer Systems models 10, 20 and 30, which were priced 20 percent below IBM's machines and offered six times as many peripheral equipment options. The new system fueled Wang's 98 percent earnings comeback in 1976. Wang did the same thing in the late 1970s in word processing, coming out with a typewriter with a television-like screen to display the printed page. The machine also incorporated an Intel microprocessor, which allowed it to be fast and small. As *The New York Times* noted in 1980, "Most of the industry now does it Wang's way."* Wang was betting that IBM's response would be weak. "With their large base of rental equipment, IBM will be reluctant to make any major improvements in their equipment very quickly," said An Wang in 1976.† Wang was right for a while. It took IBM several years to catch up, but then in 1980 it began to move ahead at the low end of the market with its Displaywriter, which was much cheaper than Wang's equivalent machine.

Polaroid is another company that knows how to retaliate. By linking price cuts to technological improvements in its film quality and cameras, Polaroid has fended off attacks by Eastman Kodak.

Xerox is another. Many people remember how Xerox lost the battle for low-speed copiers but forget how earlier it had successfully defended the top end of the market from attacks by both IBM and Eastman Kodak. Initially IBM and Kodak stole customers, but both soon discovered that copiers are complicated machines to make and operate reliably. Xerox staved off the attacks through aggressive pricing and superior engineering and by paying closer attention to its main accounts.

Aggressive price cutting by well-financed defenders is particularly effective if attackers are deep in debt and facing hefty interest charges as part of their product development. Even divisions of corporate giants like RCA and General Electric can find funding an attack a problem if their parent company is involved simultaneously in a number of different product attacks and is forced to watch the pennies.

* "Wang Labs, Head to Head against IBM," *The New York Times,* February 24, 1980, p. F3.
† "Winning With Wang," *Forbes,* October 15, 1976, p. 104.

If he is aware of a coming attack, the defender can drop his price below the cash costs of the attacker. This period of attacker vulnerability is typically quite short (say a year or so) and defenders must act swiftly. To recoup, the attacker must either draw down reserves, go to the bank, or raise equity. The latter two possibilities are particularly unappealing when losing money.

ATTACKER HUBRIS

However dangerous they look, attackers are not necessarily paragons of commercial virtue, always well informed, competitively oriented and operating flawlessly. Attackers often err by trying to incorporate too much new technology in the products, introducing substantial risk that something will go wrong during development or introduction. Something that will cost lots of money to fix. Gene Amdahl's Trilogy tried to do this with wafer scale technology. It was too ambitious, too much new technology, and has failed for the time being.

Premature market entry is another mistake attackers make. Perhaps they have pursued their knowledge building work too quickly. Perhaps they have put too much money into development and feel a need to see a return. Perhaps they have been fiddling with the technology for so long that management says, "That is it! We have had enough. We are going to market." Whatever the causes, premature entries often lose. They get crushed by a strong defensive counterattack that usually begins with collapsing prices.

Sometimes defenders get lucky and don't have to do much at all to ward off attackers. Consider the attempt in the mid-1950s by oil and chemical companies to improve the world's food supplies. When all the savants believed the number one problem facing the world wasn't a shortage of oil, then at $1.50 a barrel, but a lack of food for the developing countries, British Petroleum and Exxon (then Esso) had an idea. Their plan was to discover a microorganism—a bug—that would eat oil, multiply and become a cheap source of protein. Once the bugs had multiplied themselves to exhaustion, they would be filtered from the remaining oil, washed down, and sold to farmers as animal feed. By the mid-1960s BP was claiming its technology could be applied in refineries to produce 40 billion pounds a year of pure protein. By the late 1960s there were over 20 companies around the world doing work in this area. Today it's a different story. Total production is probably not more than 500 million pounds.

Two problems slowed the expected growth. A defender counterattacked. Productivity of soybean suppliers, the defender, soared. Even more devastating to the attackers was the price increase in their own product. Today oil is just too costly to use as a feedstock for bugs, except perhaps under special circumstances. One producer, Britain's Imperial Chemical Industries, has entered into a special deal with the Soviet Union to exploit its technology. But in the rest of the world under truly competitive marketing conditions, what looked to be a great idea almost thirty years ago has been set back largely by events that were outside the control of the companies developing the technology.

_____ AVOIDING TROUBLE

Since its founding in 1851, Corning Glass has survived and thrived better than most companies dating back that far.

Chairman Jamie Houghton and the head of R&D and Vice Chairman Tom MacAvoy (who was the former president of Corning) told me that the secret to their success was recognizing their limitations and seeking joint ventures. When it comes to picking joint venture partners, Corning has a good fix, or can get a good fix, on just about anybody.

"People selection has been the key to our joint ventures," Houghton says. "We knew what we had, and what we did not have, and we went out and got what we did not have. We found the best people we could. Set up the company. Gave it a separate board and let it go. We have had Owens-Corning Fiberglass, Dow-Corning silicones, and Genencor ceramic substrates for biotechnology as well as a half dozen others. We have been well satisfied with all of them."

Some companies try to buy their way out of trouble by acquiring technology through a licensing agreement. If all else fails, that's possible, but there are risks. There is no assurance the technology will be there when it is needed. Other competitors may have already grabbed the licensing rights. If it is available, they may be barred from buying on anti-trust grounds, or the technology may not be competitive or work as advertised. Obviously companies that stay in the forefront constantly scan the horizon to see which purchases make sense. But they augment their own internal development efforts with licensing, not the other way around.

In his book *Triad Power* Ken Ohmae strongly advocates the use of consortia and international alliances to access and sell technology. A company cannot stay up on all the proliferating technologies in some fields, he warns. But he also warns against overrelying on competitors and underinvesting in a vital function such as R&D.

Some companies have tried to invest in a portfolio of

technologies, to hedge against risk. If I buy 100 or 1,000 shares of IBM, the ratio of my return to my investment is the same. But if I put up $1 million or $10 million to develop an emerging technology my risk and my risk-reward ratio certainly are not the same. The more I invest, the better my chances get unless I'm just spending more to catch up for past mistakes. My productivity goes down, but my chances of doing the job right, of having a success, go up. Eventually, a company has to place its bets on a few key technologies. Spreading out its investment on a broad portfolio of technologies won't work if the attempt is only to spread the risk. Progress slows to a crawl.

At the core of the most far-reaching technological discontinuities is a change in the skill base of the corporation. These skill changes take a long time to accomplish, not only because it is necessary to find and train people but also to do something with current employees. It may not be possible to turn the people who make sailing ships, carriages and machine tools into builders of steamships, cars and computer-integrated manufacturing systems. It is getting increasingly difficult to bury this problem. In Europe it is "the" problem —less so in the United States and perhaps in Japan—but to neglect it would be shortsighted. The key is to anticipate the need for change in order to allow enough time to draw down the present work force through retirement and retraining.

_____ HOW TO FIND NEW SKILLS

But management also has to face the tough question of how to effectively acquire new skills. Particularly the first few people. Why would a very strong ceramist join a plastics company? Or a first-rate electronics engineer join a paper company? The best people seek their peers. They want to be where the talent is so they can do the best work. They do not want to be buried in bureaucratic battles.

One approach to this problem is to bring in a top-notch leader from the outside to head the organization. Then others will follow. Many companies that have crossed discontinuities have done this, often by filling positions with academics. Monsanto brought in Howard Schneiderman from the University of California at Irvine. Radiation Inc., the predecessor of Harris, brought in Chairman-to-be Joseph Boyd from the University of Michigan where he was director of the university's Institute of Science and Technology. Gould has outside academics on the boards of its subsidiaries. IBM brought in Lewis Branscomb from the National Bureau of Standards to be their chief scientist (as they had brought in Manny Piore from the Office of Naval Research before Branscomb). Motorola brought in Alfred Stein from Texas Instruments to head its solid-state electronics business.

Another way to bring new skills into the organization is through purchasing arrangements. Often an emerging new competitor can be converted into a supplier. The legal briefs filed in the Telex suit against IBM show that IBM was keenly aware of most small computer mainframe and peripheral manufacturers, even in regional markets, and adjusted its strategy in order to use them as suppliers.

In some businesses there are so many potentially relevant new technologies that companies will have to try to acquire them by buying their developers—turning the attacker into an ally. Many large pharmaceutical and chemical companies are continually evaluating the 200 or so biotechnology firms that have recently appeared. Whether they are correctly assessing the potential of some of these companies is hard to know, but they are beginning to make acquisitions. Eli Lilly, for example, has just acquired Hybritech (for ten times sales!), which makes monoclonal antibodies for use in medical diagnosis and production of new pharmaceuticals.

Acquisitions can be the quickest way to bring new skills into a company, but they are an expensive route toward a new technology. Buying them often requires spinning off another division to provide enough cash, and this can be painful, particularly if misperceived by the stock market.

_____ *"THE VALLEY OF DEATH"*

Once acquired, the new skill base is usually difficult to manage. The trap to watch out for is what Bob Pry, formerly vice chairman and technical director at Gould, called the "Valley of Death."

"Retention of the original entrepreneurial management in all of these companies that had grown and loved to be independent is also absolutely required. But there is a problem here; it's what I call the 3-year valley of death of acquisitions. I'm sure you can conjure up a number of examples that will fit the mold. First thing is that an acquisition is sort of like a marriage in the old days when you didn't go to bed before you were married and as a consequence you never knew quite what you were getting. And so the first thing you do in an acquisition is to replace their controller with your controller. And you send him out with instructions. He's usually a relatively young guy who is in a small division and now has all kinds of great ambitions so you send him out to see what he can find out. And of course all companies have small changes in accounting practices which they ought to clear up someday that they never really have gotten around to—either underreserved or overreserved or inventories that should have been written off some time ago—a wide variety of things that you might think of sort of stuffed under the balance sheet somewhere that you'll get around to someday. Well of course the controller can find these out pretty quick, and he does, within the first few months. Does he tell the president of that division? Of course not. He can't give him a promotion. He goes back to the CEO and the President of the major company and he tells him, 'Lookee what I found.' Well of course that makes all the entrepreneurial managers that you just made rich by stock swapping, rather furious. And generally speaking, the entrepreneurial president gets so furious he quits. Well the only person you know in that place is the controller, right? And you trust him because he just found out what was wrong with the company. So you make him the general manager.

"Well, of course the next thing that happens is that all the functional managers, the engineering, manufacturing, marketing guys recognize that this guy doesn't know doodley about the business. And they wander about for a while, but without any leadership. The sales start to fall off and things get in trouble, so they just all drift away either with the former manager, or off to do their own startups.

"So now you're left with the controller and nobody that knows anything about the business and he sort of casts around for somebody to get going and the first thing you really find out about this is that your sales have really gone to pot, you have no new products in the pipeline, manufacturing is all screwed up, and you're bailing red ink like hell wouldn't have it. So obviously, you've got to get rid of the controller, right?

"And you go out and you steal somebody good from your competitors to run the business. And you put him in and he first brings his new team in—this is about 18 months down the stream—and he then starts to go back to business with a whole new set of product lines, he has got to redo the factory, he has got to change from either direct sales to a distributor or vice versa, depending upon what you had, and by the time the fur has stopped flying you're about three years downstream with a whole new product line, new plants, new distribution, and you're beginning to make money again. That, of course, is what I call the 3-year valley of death. Now the way you get around that is relatively simple but hard to follow. When you bring your new company on board, you leave them alone. You literally leave them alone. If they have a board, let them have their board. Now you recognize that what you've done with the president is stock swap with him. He is used to his appreciation of stock being at least 50 percent per year, right? Now he's in a company that grows maybe by a few percent per year. So about six months downstream he takes a look at his portfolio and says, 'Holy cats, I'm not making any money.' Next thing you know he appears at corporate headquarters saying what can I do to make this dumb company get off of its dead can and start growing again? You got it. Because now he has appeared and is interested in the way things operate generally. It's very difficult.

"All of us are alike. Let's get in there and find out what's doing, right? Corporate is here to help you. Anybody that has been from a division understands those words pretty well." *

Avoiding Bob Pry's three-year valley of death is crucial in acquisitions. Whether it represents an easier way to change the skill base of a company than the patient nurturing of

* Pry, Robert, "Organizing for Technology Implementation," speech delivered at the Fall Meeting of the Industrial Research Institute in Orlando, Florida, November 1984.

newly employed scientists and engineers convening for a newly recruited technology head like Howard Schneiderman at Monsanto depends on the patience of top management and the marketplace.

_____ *SEPARATE FRONTS*

The evidence suggests that the attacking and defending ought to be done by separate organizations. Practically all the requirements for defenders and attackers differ. The key strategy for an attacker is probably based on technology whereas the defender's strategy may require superior marketing. The pace of decision making for an attacker often needs to be faster than that of the defender. The size of the attacker's organization thus needs to be smaller than the defender's, permitting a wholly different set of managerial controls to be developed. The skill base for the attacker is often quite different from the defender's. These two sets of requirements can't be mixed. If they are, development time slows to a crawl as supposedly optimal decisions are made.

Yet this is often what happens because the attack strategy is not seen as a primary but almost as a backup strategy for the defense if it fails! NCR's comment in 1971 that computers were "an important *adjunct* to cash registers" or American Viscose's comment in 1968 that "we still sincerely believe that rayon is the best tire cord" are cases in point. The planning systems at many major corporations allocate capital to the defense of existing business strategies and give only what is left over to attacking strategies or new opportunities.

This happens because a defender, even if he recognizes the potential for a discontinuity, assumes he can control the pace of the transition from the old technology to the new. Having separate organizations for attack and for defense does not make sense because there doesn't seem to be any real need to hurry. Many CEOs have said, "Why should I have a laboratory on the West Coast with one overhead structure, and another on the East Coast to do it a different way with another overhead structure? I could have one laboratory in St. Louis, with a single overhead and get double the leverage out of those people." That is economically true, but effectiveness is the issue, not efficiency. Reaching for greater efficiency when there is a much more effective option available is suicide. Inevitably it pushes a company toward maintaining the current technology and makes it more vulnerable to competitors who do not have two technologies to contend with.

The best organization to exploit current technology is functional. Independent technology, manufacturing, marketing and service departments optimize each of their functions to produce a product. What happens when you bring a new technology into this arrangement? Where do you put it in the organization? If you bring in new engineers do you put them in the current engineering department? If this little group succeeds in designing a new product, do you then assign another group in your manufacturing organization to make it (when it has a chance) and yet another group to market it (along with established products)? How do these different groups coordinate and make the quick decisions necessary to compete in a new market? They do not. The strong lines to the top go through the functional side. If a company wants to be competitive with a smaller rival that is designing, making, and selling only the new technology, then it must spin that new technology group out from under its defensive wing. This is precisely what IBM did by setting up a separate organization in Boca Raton to develop the PC. It is what Boeing did to develop the 757/67. And what Xerox did when it set up its Systems Group to develop noncopier office products.

It does create problems. How do you motivate the personnel in the defender group? Which do you select to head the defender group—a rising star or a mature senior executive? How do you compensate the heads of the attacking and defending businesses? Equally or according to assets controlled or "strategic importance"? The answers to these questions are not often (nor should they be) consistent and add up to messy solutions. But more often than not the neat solution, the integrated solution, is less effective.

GAINING A TECHNOLOGICAL EDGE

The key to both attack and defense is picking the right technology. Identifying the alternatives and recognizing their limits does not usually lead to an obvious answer. The best strategy may be to pick the adolescent technology with the highest limit because it is never clear when a new technology is really going to emerge. Go for the highest potential technology because customers almost always want more performance when they can get it.

Early on, through alliances, licenses and in-house research, the best approach may be to pursue several possible technologies. Dartmouth's James Brian Quinn summarized a two-and-a-half-year study of innovation by concluding that, "Many of the best concepts and solutions came from projects partly hidden or 'bootlegged' by the organization. Most successful managers permitted significant chaos and replication in early investigations but insisted on much more formal planning and controls at expensive later stages." Quinn is right but makes another important point, that eventually we need to make an informed but not totally rational choice about which technologies to try to push up the curve. "Choosing which projects to kill is the hardest decision, the essential tension, and the art of large-scale innovation management. Repeatedly, successful managers have said, 'Anyone who thinks he can quantify this decision is either a liar or a fool . . . There are too many unknowables, variables . . . Ultimately one must use intuition, a complex feeling, calibrated by experience . . . It's a judgment about people, commitment, and probabilities . . . You don't dare use milestones too rigidly.' " *

Patents are the traditional way to get or protect a technology. Unfortunately, most corporations are skilled in getting around patents. In many cases patents become important only as tradeable assets. They are the ante for getting in the game. If you have a large patent portfolio, as do companies like Bell Labs, IBM, Du Pont, or Hewlett-Packard, it's possible to trade patents like baseball cards until you get the set you want. It is better to have them than not, but patents rarely are strong enough by themselves to prevent leapfrogging.

Successful defenders seem to be guided by a vision about how the battle will go over the long run. They go into the first encounter with a sense of what the end game will be like. Will it be a battle for low cost? For control of key distribution channels? For the key product design? This point of view then becomes central to the conceptualization of the

---

* Quinn, James Brian, "Large Scale Innovation: Managing Chaos," *Tuck Today,* June 1985, pp. 2–7.

strategy for winning the first battle. As a result, they know how to judge each piece of news. They can stand above the battle and still be in it.

Above all, successful defenders are wary. They understand the toughest challenges are not likely to come from familiar quarters but from unexpected attackers armed with new skills. Such challenges are threatening both because they offer a customer more than he can get for his current outlay, and because they offer something different. "Quick chill" cans, gasoline-saving tires, plain-paper copiers, "instant" pictures, home banking, no-wash diapers, and slow-release pills. All things we never had before. All things that changed the basis of competition. None of them could be thoroughly researched in advance. The only defense was to be alert and responsive.

How do companies stay alert when there are no economic signals of deterioration or, indeed, when everything is going well? For me the way to do it is to use the insights of S-curves to create some early warning indicators, some thermometers of the corporation's technological health. If there is a discontinuity coming, chances are that the current technology is close to its limits and a new approach is being tried by some upstart, nonconventional competitors.

There are several questions that need to be thought through:

1. *Is there increasing discomfort among top management about R&D output?* There may be a visceral feeling that the business is going fine but the scientists in the lab aren't as productive as they once were. Most companies don't pay enough attention to this. The CEO doesn't have enough self-confidence to say, "I'm uncomfortable, and therefore I'd better

look into this thing." Instead he says, "I'm uncomfortable, but those technical guys know what they're doing." Top management discomfort with what's going on in the lab may be important early warning signals.

2. *Have development costs and delays started increasing instead of falling?* As a technology gets closer to the top of its S-curve and closer to its performance limits, it takes greater effort to produce even small changes in product or process performance. Small changes in performance specifications can require large changes in the resources required. I know of no resource estimation system that takes this into account at present. The general assumption that the next product will be cheaper than the last to develop because we have traveled along the learning curve often conflicts with reality. Because of diminishing returns, it will get more and more expensive.

3. *Are you doing more process R&D, less product R&D?* This is a natural shift in emphasis, but it's also a sign of technological maturity and needs to be recognized as such. There's not much that can be done save switching to a new technology.

4. *Is creativity waning?* Are measures of creativity—patents applied for, new products developed, significant process innovations and new ideas—falling off? How could they do otherwise unless the basic technical path is switched from the top of one S-curve to the bottom of another?

5. *Is there disharmony and discouragement in the labs?* As the technology matures, you find lots of disharmony in the laboratory or at least you can sense a lack of excitement. Lewis Thomas, president of the

Sloan-Kettering Cancer Institute, describes this in his book, *Lives of a Cell*. ". . . a good way to tell how the work is going is to listen in the corridors. If you hear the word, 'Impossible!' spoken as an expletive, followed by laughter, you will know that someone's orderly research plan is coming along nicely." * He's right. In a one-hour tour I can tell more about R&D productivity than from any discussion with a lab director. You can feel the mood.

6. *Is market segmentation becoming the key to sales increases?* Segmentation can bring more profit and is the right tactic for many companies. But segmenting the market doesn't address the underlying technological health of the business. It indicates you can no longer get "across the board" improvements in your competitive position through major advances in the technology and probably indicates a technology near the top of its S-curve.

7. *Are there wide differences in R&D spending among competitors with no apparent market effects?* After merging, two oil companies were seeking ways to become more effective. One executive saw that the other was spending several times as much on gasoline R&D for no discernible technological or profit advantage. He realized both companies were at about the same point on the S-curve of refining technology. Spending $5 million, $50 million, or $500 million on the technology didn't make any difference. If you're at the limit, you're at the limit.

8. *Have there been frequent changes in R&D management with no impact?* Is there a constant hiring and

* Thomas, Lewis, *Lives of a Cell*, New York: Viking, 1974.

firing of the chief technical officer? This happens when the CEO loses patience with technology development but keeps making the same error. Each time he makes a change he puts the new guy in the same box and under the same restraints as his predecessor.

9. *Are some leaders losing share to smaller competitors in market niches?* This is a sure signal that there may be a new technical approach afoot. Remember discontinuities require at least two S-curves. Small competitors with limited resources can outpace large rivals because the dollars they put into their technology result in much more progress. Being on a different S-curve (integrated circuits), tiny TI could take on and clobber giant Westinghouse. The same thing will probably happen in pharmaceuticals, where more than half of the advances will come from biotechnology even though today biotechnology probably gets less than 10 percent of the industry's total R&D funding. Higher R&D productivity always wins the day.

10. *Are supposedly weaker competitors succeeding with radical approaches that everyone else says cannot work?* As we saw earlier, Boeing was a failing company when it opened the age of commercial jets by radically altering the design approach of the British Comet. Often the radical moves come from weak competitors. There isn't much else for them to do except close down.

Being more alert no doubt is due in part to extensive efforts to detect discontinuities. But perhaps more important than the mechanics of surveillance is the belief that discontinuities exist and represent an opportunity for changing competitive leadership—that it is imperative to move to the attack even as you put up the best possible defense. Companies that manage to access discontinuities seem to believe, almost to take as a matter of faith, that if there is room to move technically, then leadership can be sustained only through continual change.

Everything is oriented toward change. P&G insists, for example, that every employee have a set of objectives not only for his current performance but for how he is going to change the way he does things. IBM's strategy and culture are based on change. Few corporations reorganize as frequently as IBM. These reorganizations are partial rather than total, but the emphasis is always on change. Robert Galvin, chairman of Motorola, comments, "My father said that if you keep purposefully in motion, eventually you will get something done."*

Just recently I was discussing some of these ideas about discontinuities with the executives at a leading West Coast biotechnology company. I had looked forward to this meeting because mostly I spend most of my time talking to defenders, helping them see their situation in a different light. Here was a chance to spend some time with an aggressive attacker. Now I could really test my ideas.

After the formal part of the talk, several of us broke for lunch. I asked the president whether he felt these ideas applied to him. "Oh yes," he said, "those are our problems exactly, and your suggestions for getting out of the trap were helpful too."

Puzzled, I said, "What do you mean problems? It looks to me as if you have got nothing but opportunities. You are trouble for somebody else."

---

* "There's a New Team at Motorola," *Business Week*, April 23, 1966, p. 111.

"Well," he said, "I am glad to hear you say that, but from my point of view we are the defender. We made the first big breakthrough in this business. Well, not really the first. Our friends south of here really cracked the market, but they got too big, too fast. We capitalized on the rapidly building knowledge base, and eclipsed them. In that case we were the attacker. But now we are in their position. We are working like mad trying to figure out how to protect our technology from some of the new companies in the field."

"New" companies indeed! His firm is barely three years old. Already the attacker is taking on the gray-haired looks and worries of a defender, fearful he will be eclipsed by an even newer, more cost effective technology than the one he is pursuing.

But isn't that the way it is in business today? Events and changes come at a swifter and swifter pace. In the biotechnology industry three years is a lifetime and no sooner do we get to know today's heroes than they become tomorrow's forgotten men. It happens in other industries too. At the end of the 1970s, Airbus ruled the skies. Today Airbus is struggling for its life. Overnight attackers become defenders and if they want to survive they must learn a whole new set of skills. They must believe in and discern the structure and pattern of technological competition. Otherwise, the skirmishes take place, battles are won and lost, and they are no wiser. Above all, they must learn the art of being both defender and attacker at the same time. This is the hallmark of successful companies that seem to persist through battle after battle.

NINE

PHOENIX: LEADERS WHO STAY LEADERS

Our thinking about growth and decay is dominated by the image of a single life-span, animal or vegetable. Seedling, full flower and death. "The flower that once has blown forever dies." But for an ever-renewing society the appropriate image is a total garden, a balanced aquarium or other ecological system. Some things are being born, other things are flourishing, still other things are dying—but the system lives on.

—John W. Gardner,
Self-Renewal, 1981

There are not many companies in the United States or around the world that appear to be able to manage through technological discontinuities. But if you were naming names you would likely include Gould, P&G, Corning, Harris, Monsanto, IBM, Merrill Lynch and Citibank among them.

Gould's 1983 annual report announced: "Gould in 1984 will complete one of the most substantial restructurings in corporate history, dramatically changing from diversified industrial products to a total electronics company." And indeed the record supports the claim. In 1968, when Bill Ylvisaker took over at Gould, it was a battery and engine parts company with sales of $124 million and earnings of $4.2 million. In 1984, sales were $1.5 billion (average sales growth of 17 percent) and earnings were $86 million (average earnings growth rate of 21 percent). The Value Line investment advisory service expects sales to be in the $3 billion range and earnings to be about $220 million from 1987 to 1989. Moreover, Gould no longer makes battery and auto parts, but is attacking in electronics with products like gallium arsenide chips, medical products based on electronics (such as invasive blood pressure measuring systems), 32-bit microprocessors (32 bits brings much higher computing speeds than either the present 8-bit or 16-bit systems), imaging and graphics software, which is a key to automated design, and factory automation. Had Gould stayed in batteries and engine parts it would undoubtedly be only a shadow of its present self.

In fact, it has undergone two metamorphoses since Bill Ylvisaker took over. The first came as soon as he moved in, when Gould acquired Clevite, the ex-germanium-based semiconductor company. Clevite also made batteries and other industrial goods. Ylvisaker paid a hefty 20 times earnings for the strategically weak company, but it did move him into electronics, which is where he wanted to be. In 1977, Gould acquired the $503 million ITE Imperial Corporation, which made circuit breakers and other electrical distribution products. The first "new Gould" took shape, as electrical products augmented Gould's battery line. But things did not work out as expected with ITE. The price of oil soared and utilities, which were a prime customer for ITE, cut back their purchases. Margins and returns began to slide. Profits dropped from $101 million in 1978 to $73 million in 1980. Efforts to control the business did not work. Executives left, including the president.

At this point, Bill Ylvisaker went for a second metamorphosis. As the British journal *Management Today* said, "The following year (1981) Bill Ylvisaker did a strange, remarkable thing. He tore it all down." Strange to the British perhaps, but not to the U.S. investment community, who saw the move as "nothing short of a sensational stroke. To have perceived the need for the redeployment in the first place was insightful. To have had the nerve to commit to such a course was courageous; and to get it executed was perceived by Wall Street as a master stroke." *

Courageous indeed. Gould sold its $458 million Industrial Group (defender products) for $375 million as well as additional parts of the company that account for sales of another $102 million. Sales dropped from $2.2 billion in 1980 to $1.6 billion in 1983, and earnings stayed essentially flat, but cash was now on hand, and debt/capital dropped from 41 percent in 1979 to 26 percent in 1983.

Gould then built a solid position in fast-moving industries through its acquisitions of Systems Engineering Labs, which makes super minicomputers; American Microsystems, which makes a range of application-specific integrated circuits; DeAnza Systems, which makes computer-based high-resolution image processing and display systems; SRL Medical, which makes computer-based cardiopulmonary diagnostic systems; Compion, which designed software for use with AT&T high-performance UNIX operating systems; and Dexcil, which makes gallium arsenide field effect transistors. (He also tried for Mostek and Fairchild Camera and Instrument, but lost to Harry Gray at United Technologies and to Schlumburger.) † And not too soon. Just after Gould got out of some of the more mature defender products (high-voltage switching equipment, for example), they went through hard times.

The stock market, however, did not apparently view all this as a "master stroke" just yet. The stock price remained essentially flat at about $30/share. But by late 1984, *The Wall Street Journal* reported, "The strategy must be working,

---

* "Gould's Golden Gamble," *Management Today*, February 1984.
† Probably much to his pleasure at this point since both these acquisitions have been failures.

because many analysts have turned bullish on the company." * Value Line expects the stock to increase to the $70 to $100 range by 1987–89. Indeed, Roy Chapin, former chairman of American Motors, said that the metamorphosis of Gould "is almost a minor industrial miracle." It is possible to transform yourself. But it takes more than buying and selling companies.

_____ *ATTACK YOURSELF*

On March 1, 1985, Wesley Thompson, recruited only 14 months before, resigned from Beatrice. As *The Wall Street Journal* said, "Mr. Thompson's resignation underscores acute problems at Tropicana. The orange juice market has become more competitive with the entry of Procter & Gamble's Citrus Hill brand, and the increased marketing efforts of Coca-Cola's Minute Maid. The industry has been ripe with heavy marketing spending on the one hand, and price cutting on the other. In the past six months, Beatrice has cited Tropicana as one of the major reasons for the drop in operating profits. James L. Dutt, Beatrice chairman, has said repeatedly that Beatrice will stand behind Tropicana and will not give up in its fight for market share, even though the bottom line may suffer in the short run. In the past five months, however, Tropicana has been losing ground to Minute Maid (Coke's product) and Citrus Hill (P&G's product)." †

In fact, P&G's move into Citrus Hill orange juice through its acquisition of Ben Hill Griffin in 1982 is just

* John Bussey, *The Wall Street Journal*, October 3, 1984, p. 1.
† Morris, B., "Beatrice Co. Says Presidents of Two Units Quit," *The Wall Street Journal*, March 1, 1985.

another example of how P&G goes on the attack. The move into juice drinks was made only after many years of R&D in the beverage business, in both coffee and soft drinks. It is reported that the initial Citrus Hill brand tasted no different from Minute Maid or Tropicana. But *Business Week* speculates, "The rollout of a conventional 'me-too' brand could be a P&G ploy to draw out rivals and strategies, drain their bank books, and be followed by the relaunch of the improved Citrus Hill next year." * P&G has a patent on a better-tasting product (Citrus Hill Select) that is now in regional rollout. The point is that big companies in conventional "low-tech" products can attack too.

An earlier example is P&G's move into paper products. P&G perceived an opportunity in paper products in the 1950s when they saw a large potential for paper to substitute for cloth. Further, current products had uneven quality and were made by sleepy competitors. P&G then began to strengthen its knowledge base in wood, cotton linter pulp, and paper to develop a fundamentally new paper manufacturing process. It was here that it developed the rudimentary concept of Pampers.

P&G figured it needed to augment its own expertise in paper manufacturing with someone who knew the consumer tissue market and competition, and which had facilities to make products. So it acquired the Charmin Paper Company in 1957.

It proceeded to integrate the Charmin people into P&G slowly. It purchased timberlands, developed a paper products sales force, and selected a new advertising agency.

Not satisfied, they continued to develop the Pampers process through development of a new drier which helped them

---

* "P&G Dives Into Orange Juice With a Big Splash," *Business Week,* October 31, 1984.

improve the softness and absorbency of their products. In 1967, when they were ready, they launched Charmin and White Cloud toilet tissue and Puffs facial tissue, and tested Pampers. In 1970 they rolled out Pampers into the marketplace and the rest is history. P&G now has about 60 percent of the multibillion dollar disposable diaper market. Personal care, a new segment for P&G, has grown from $625 million in 1970 to $4.8 billion in 1984, a 16 percent per year growth rate. It accounted for 38 percent of P&G's sales and 44 percent of P&G's pretax earnings in 1983. It's a good example of how a big company can make big money by attacking.

The strategy of attack and constant churn is not new at P&G. Nor is the concept of simultaneous attack and defense, i.e., attacking your own products. The origins can be traced back to the 1930s and a then young man destined to become the chairman of P&G, Neil McElroy. McElroy is considered to be the inventor of the "brand management" concept. As Oscar Schigall, who chronicled P&G, tells the story:

"It was P&G's introduction of Camay which triggered the brand management approach McElroy proposed. For Camay's performance had become disappointing. In the beginning it was not the great challenger of Ivory. Why? Company executives decided it was being held back by 'too much Ivory thinking,' particularly in the advertising agency which handled both brands. They felt that Camay's advertising had been weakened because of possible reflections on Ivory. In effect, Camay was not being allowed to compete freely with Ivory. It was the victim of an obvious conflict of interests.

". . . the new agency selected for Camay was Pedlar & Ryan, New York. They were assured there would be no restrictions on competition. Henceforth Camay and Ivory would have to fight each other for a place in the market.

Camay would be free to advertise as vigorously against Ivory as against other companies' Lux, Palmolive, and Cashmere-Bouquet." *Time* called the idea a "free-for-all among P&G brands, with no holds barred."

A detractor said it was like starting fights in a family. Schigall writes, "McElroy responded that if the brands of competition like Colgate-Palmolive, Lever Bros., and others had been unable to destroy Procter & Gamble products, why fear homegrown competition? ... The brands would vie with one another like brothers in a race, not like enemies. Internal competition, he maintained, would bring into play every talent, every ability, every tool possessed by brand managers.

"The *Time* story concluded: 'Eventually McElroy won his point. He persuaded his elders that the way to keep fast-growing P&G from becoming too clumsy was indeed to have it compete with itself.' It was a concept new to American industry. Never before had any American firm encouraged such competition among its own brands." *

In essence, P&G's present strategy has three parts to it:

1. Vigorously defend existing business through aggressive marketing and cost containment.
2. Invest in a selection of small businesses in areas that might attack present businesses. Watch the portfolio carefully. Pick the most successful ones.
3. Pursue the attack strategy internally through the brand management system with dedicated market and technical resources. Transfer the methods of the successful small companies to the larger brand.

* Schigall, Oscar, *Eyes on Tomorrow,* Chicago: Doubleday/Ferguson Publishing Co., 1981, pp. 160–64.

The ability to be both attacker and defender has produced exceptional results for P&G over the past 40 years.

JOINT VENTURES

Corning Glass would probably not be on most people's lists of companies that have continuously rejuvenated themselves. But that's exactly what it has done since it was started in 1851. Today it is a $1.5 billion company in ceramics, health science, and telecommunications. In addition, it owns a substantial portion of companies like Dow-Corning and Owens-Corning Fiberglass, which have sales of another $3 billion or so and provide Corning with a stream of dividend income.

Jamie Houghton, descendant of the founders and recently appointed chairman, traced several major changes in the company's history. In the beginning, Corning was a maker of specialty glass—tableware, decorative glass, pharmaceutical glass, and railway lanterns. Then in 1879, Corning teamed up with GE's founder, Thomas Edison, and made his experimental incandescent light bulb. Edison, it is said, picked Corning because of their reputation as a research-based company. By 1900 the company was primarily a lighting company with some specialty glasses.

In 1908, Corning established a research lab, one of the earliest in American industry organized for basic research.* After a long period of trial and error, the lab came up with Pyrex and suddenly Corning was into consumer goods.

The lab produced other significant results. For example, one of Corning's most important innovations was the ribbon

---

* Houghton, J. R., "The Role of Technology in Restructuring a Company," *Research Management,* November/December, 1983.

machine. In a ribbon machine, a ribbon of hot glass sags through holes in a continuous moving belt while compressed air from below shapes the sagging glass into a bulb shape. A ribbon machine today can produce 2,000 bulbs a minute. It is the machine that put the electric light within everybody's economic reach.

Corning kept up the pace by further leveraging its technologies with joint ventures and acquisitions. For example, another glass-making process to come out of the labs, the "updraw process," pulls a cylinder of glass from a pot of hot glass. The updraw process made possible the mass production of tubing in different diameters—some as small as thermometers. From this technology came the ability to "spin" glass, that is, to produce it in fiber form. The resulting product—fiberglass—finds a wide range of uses today in the construction industry. The first commercial product came in 1935; Owens-Corning, a joint venture with Owens-Illinois, followed in 1938.

In 1930, Corning was experimenting with silica (the raw material of glass) and plastics. The result, in 1936, was silicon, which has all sorts of industrial and medical uses today such as breast implants. Dow Chemical had the manufacturing and marketing base for the product, so Dow-Corning was formed.

Work for the Department of Defense to produce radar bulbs for the war improved skills in bulb formation. From this base centrifugal casting was developed. That brought the world TV tubes. And that took Corning into electronics. Signetics was added to move further in that direction, but the racy electronic components business proved too much for Corning, and it sold Signetics to North American Philips in the late 1970s.

Nevertheless, Corning persisted in electronics and has

found its niche with ceramics for capacitors and resistors. This and Corning's strong position in optical wave guide material have given it a position in telecommunications. The wave guide research began at Corning but was augmented with considerable know-how from Bell Labs.

And Corning has done the same thing in medical products, building on innovations in porous glass to make porous ceramic substrates for enzymes. This put Corning into biotechnology, medical instruments, medical diagnostics, and medical services. Here they have done a joint venture with biotech leader Genentech and have also acquired Gilford Laboratories, Met Path (clinical labs), and Kansas City Biologicals (cell growth medium).

Of course, Corning has not had a perfect record. For example, they are still a major manufacturer of ophthalmic glass for eyeglasses and have missed the plastic lens market, which has been substantial.

The net result of all of these transformations, all based on attacking early while managing defending business well, has poised Corning for growth. Stock analysts predict earnings will be up 100 percent to $200 million by 1987–89, and they feel that stock price will double too. And all this in sleepy old glass.

_____ ONGOING METAMORPHOSIS

Harris Corporation is a $2 billion electronics and communications company in Melbourne, Florida. It is a market and technology leader in most of its product lines. But it was not always that way. In 1967, Harris was Harris Intertype, a Cleveland-based producer of mechanical printing presses, although it did have a small business in manufacturing radio and TV station transmitters. Harris has been through what *Forbes* called in 1980 a "remarkable metamorphosis."

> The stage for today's growth was set back in 1967 when the company was Cleveland-based and known as Harris Intertype. Its chairman, George S. Dively, then 64, had built Harris from under $10 million in sales to $195 million. Looking ahead, Dively correctly saw the printed word and spoken word merging in electronics and knew he had to pick up electronic know-how for his then mechanical printing presses.
>
> Dively and his heir apparent, President Richard B. Tullis, paid $56 million to acquire Florida-based Radiation Inc., a high-technology space and defense company.*

The cost of George Dively's vision was high. The acquisition price of Radiation was itself beyond what the pure finances would have suggested it was worth, and the toll of introducing a constant stream of new products cut into earnings, which remained essentially flat from 1968 through 1975.

Then things exploded. Sales went from $479 million in 1975 to $1.55 billion in 1981 (a 22 percent growth rate) and earnings went from $0.9 million to $91.8 million in the same period. Even though 1975 was a particularly bad year for earnings, it is a terrific record.

Harris initially took its new electronics skills and put them into the printing press rooms of newspapers around the

---

* Flaherty, R. J., "Harris Corp.'s Remarkable Metamorphosis," *Forbes,* May 26, 1980, p. 45.

country in an attempt to replace typewriters with video terminals. In addition, Harris improved the facsimile transmission of wire photos. In short, it attacked its old heartland with the new technology.

But printing was not neglected. In 1980 it was reported that Harris had trouble keeping up with the orders for its $250,000 to $5,000,000 web offset presses, which replaced old letter presses.

As it continued to grow, it used its experience with defense systems to build more into communications technology. In 1980, Harris merged with Farinon, a producer of high-technology telephone equipment. Harris also developed a capability in electronic components. It was tenth in integrated circuit sales in the United States in 1980. At the time, Harris claimed to be second only to Intel in profitability and well ahead of Texas Instruments, Fairchild and Motorola. The company said that since its start-up in electronics in 1962, it had had only one losing year.

All the signs are that Harris is now at the beginning of its second metamorphosis. It has tackled the office products area where IBM, Xerox and Wang seem to enjoy beating each other up. The start was again rocky, with a long delay in getting the first system out due to software problems. By the time it finally got to the market, the product did not possess clear advantages. To bolster its position, Harris bought Lanier for $280 million in stock. Lanier once had a top position in the marketing of word processing systems. To pay for Lanier, Harris sold its old defending product line, the printing press business, for $230 million in cash, and this has once again hurt performance and stock price. Value Line is positive, however, and sees sales growing to well over $3 billion and earnings growing to $190 million in the 1987–89 period. If Harris accomplishes that, it will be another example of a company successfully pursuing a strategy of simultaneous attack and defense.

_____ HINDSIGHT IS AN EXACT SCIENCE

The actions necessary to cross technological discontinuities frequently attract criticism, sometimes rightly so because they are often questionable and risky. By any conventional financial yardstick, Gould paid too much for Clevite, SEL, DeAnza, American Microsystems and Compion. Harris paid too much, according to the business press, for Radiation and Lanier and should not have sold its printing press business. As *Business Week* noted, "Boyd is sticking his neck out by selling the printing business, which the company says is the world's largest producer of printing equipment. It had contributed one quarter of the company's revenues in the fiscal year ended June 30, 1982. More important, the printing sector boasted better profitability than most of Harris' high technology business." *

When P&G installed its product manager system, it came under heavy fire. Schigall writes:

> Did the plan, once adopted, delight everybody? A former officer said, "No. There are always people who resist change, even resent it, especially old-timers who have been working under a familiar system for many years." Some of them thought the company was about to be shattered by internal disruption. Others felt their chances for advancement would be

---

* "Harris Is Raising Its Bet on the Office of the Future," *Business Week,* July 18, 1983, p. 34.

destroyed now that the traditional avenue for promotions would be supplanted by this new setup. No,
there was not what you might call universal happi-
ness. But time was to prove the doubters and dis-
senters wrong, for the reorganization of P&G
brought it greater strength than ever.

Crossing discontinuities is not pleasant, nor are all actions successful. Many employees at Gould were badly de-
moralized during the metamorphosis and many good people
left. Critics say Bill Ylvisaker was too quick on the trigger to
wait for his investments to pay off. They claim his impatience
caused the stock price to flatten for years. Harris's move into
newspaper-room video terminals was not financially successful and indeed Harris was beaten to the punch by an even
earlier attacker, Computervision. P&G has not only had
product successes, it has had failures too, like Rely, which
resulted in toxic shock. And many would say Corning is far
from an exciting company; they certainly stumbled in electronics. But despite all these human problems and implementation mistakes, these companies have done much better
because of their attempts to cross discontinuities than they
would have if they had attempted to avoid them.

Neither Gould, P&G, Corning Glass, nor Harris have
had continuously excellent records, and no doubt each of
them will continue to have some failures. But in technology
and business the goal isn't to bat a thousand. It's much like
real-life baseball. How many batters have consistently been
among the top ten? Since 1950, not many. Stan Musial held
a spot in the top ten from 1950 to 1958, and Roberto Cle-
mente did it from 1960 to 1969. The great batters like Willie
Mays, Pete Rose and Carl Yastrzemski had a good year or
two and a bad year or two and then a good year or two.

Pulsing, not continuous, success. It takes great effort, skill, and luck to remain among the top hitters in baseball or among the best companies. For the majority it's just an elusive goal. Today, P&G and Harris have hit some slumps, but that's to be expected. What counts is their resiliency. They will stay on top. Unlike aging baseball players, companies can reinvigorate and renew themselves.

TEN

LEADING
METAMORPHOSIS

One should recognize and manage
innovation as it really is—a tumultuous,
somewhat random, interactive learning
process linking a worldwide network of
knowledge sources to the subtle unpre-
dictability of customers' end uses.

—James Brian Quinn, 1985

Everybody's advice to the CEO about every function—
operations, law, planning, marketing, distribution, finance—is that he has to pay more attention to it. But if
technology is the key to continued corporate success, the case
for CEO involvement in technology is stronger than it is in
other areas. The CEO has a unique role to play either as the
architect of his company's strategy or the maestro who or-
chestrates its development. As we saw in the case of biotechnology, strategy and technology heavily influence each other.
The CEO has to understand and make decisions about which
technological paths to follow. And that means up-front involvement well before any funds are committed or positions
have hardened.

He has to do something else as well. The next ten to
twenty years will be filled with technological discontinuities.
It's not a case of counterattacking once or twice a decade,
but continuously. Being able to manage through discontinuities will be crucial for corporate survival. The CEO needs to
be intimately involved in developing the necessary management approach and culture to meet this challenge.

This does not mean that the CEO needs to be a scientist
already competent in upcoming new technologies. He does
not have to be the chief engineer, and if the company has

multiple divisions and product lines he cannot be.* There are many examples of entrepreneurs who have led the technical charge. Some were scientists: Edison of GE, Eastman of Kodak, Wilson of Xerox, Land of Polaroid. Others were not: Tom Watson, Sr., at IBM, George Dively at Harris and Bill Ylvisaker at Gould are examples of CEOs without technical backgrounds who nevertheless oversaw technological decisions which took their companies in new directions.

In large corporations with hundreds of technological options it could be a significant advantage to have as CEO a man who understands the process of scientific discovery— the principles of productivity and yield as I've described them. That may mean that we will see more managers with scientific backgrounds reach the top position. This was the case with John Reed at Citibank. Last year Merck, the largest pharmaceutical company in the U.S., appointed a 55-year-old physician and biochemist, Roy Vagelos, to its top post. Reports *Business Week*, "Recruited from academia ten years ago to run Merck's research laboratories, Vagelos has already set the company on a bold and risky course: Instead of testing chemical compounds as treatments for specific ailments, he has embarked on a campaign to learn how diseases work. Now, as CEO, he must make sure his long-term strategy turns out products that sell."† Vagelos's colleagues say that having a scientist as CEO is advantageous because he understands the ups and downs of research.

McKinsey once did a comparison of CEO backgrounds and corporate performance. The results suggested that companies led by technical or marketing people out-performed

---

* For a discussion of the role of the CEO in technology in different kinds of companies, see Maidique, M. A., "Entrepreneurs, Champions, and Technological Innovation," *Sloan Management Review*, Vol. 21, No. 2, Winter 1980, p. 59.
† "A Research Whiz Steps Up from the Lab," *Business Week*, June 24, 1985, pp. 87–88.

those run by financial people by a substantial margin. There is a bit of good news here. Management Practice Consulting Partners recently surveyed the backgrounds of new CEOs. In 1978, about a quarter came from the technical and marketing functions; by 1982 about 50 percent. Maybe we are moving in the right direction. Certainly more boards of directors ought to consider appointing technologically oriented executives to the top spots in their companies.

CULTURE AND UNDERSTANDING

Even though the CEO need not be up to date on all relevant technologies, the competitive outcome of a market battle often requires an understanding of what appears to be obscure technical details—the band gap of germanium versus silicon, the differences in molecular structure of naphthalene and orthoxylene, the composition of complex and simple sugars. These details dictate the range of options management has at its disposal. Understanding them and their possible competitive implications often depends on the atmosphere inside the corporation and people's expectations. Is open challenge of the conventional approach encouraged or discouraged? Are alternatives explored? Is the customer's viewpoint adequately represented? If the proper environment for constructive challenges is not established, then it is very unlikely that a company will embrace a new technology, let alone develop the broader capability to manage technology and discontinuities.

The man who translates the CEO's vision and balance into an R&D program is the Chief Technical Officer. The strength of his relationship with the CEO is thus important. Unfortunately this relationship is often weak. The Conference Board asked the CEOs of the 400 largest companies in the United States who were their most trusted advisors. CEOs mentioned themselves first—a positive sign. The next most frequently mentioned is the top financial officer followed by the general operating officers, then marketing, human resources, corporate counsel, planning, production, general corporate staff—and then R&D. In only one case in five was the head of R&D considered a member of top management.

Ken Ohmae says that in Japan the chief technical officer would make the list of key advisors 80 percent of the time. That would put him about third on the list, instead of eleventh. If that is the case, we have a big problem in the United States. How are we going to compete in a world full of technological potential if the executives who know most about our technology are not close to those who control how the funds and people inside our companies are deployed?

Whose fault is it? In part the CEO's, but my impression is that in many cases the R&D managers in this country are not broad enough to relate to the CEO and to understand his concerns. The R&D chiefs that seem to do the best are often those that have had direct line operating experience. Indeed, the best R&D chiefs are qualified for the CEO's job. Tom MacAvoy at Corning was president until he voluntarily stepped aside to become Vice Chairman and head of technical operations. Bennett Nussbaum at Pepsi or Len Baker at Union Carbide are other examples of R&D heads qualified to move up.

The argument often made is that the chief technical officer really has to be the best technologist in the labs or he won't get adequate respect from the scientists and engineers. That is contrary to my experience. The chief technical officer does not have to be, and in many cases where he is effective, is not the best technologist. He does not have to be the most creative scientist or the most innovative engineer. But he has to be able to translate the CEO's vision into technical terms by researching the alternatives available for pursuing that vision, finding the limits of the alternatives, the estimated costs and benefits of each alternative, and assessing the extent to which it will provide the basis for a sustainable competitive advantage. He also has to understand how best to acquire technologies and how to apply them. In short, he has to understand effectiveness and efficiency, and the difference between them. That is his job. If he does that well he will receive all the respect he can handle.

On the other hand, the CEO needs to understand a lot more about the management of technology. Otherwise, he will not raise essential questions with the lab directors about limits and alternatives. Both sides are going to have to work at this. They are going to have to find a common language that is meaningful in their company. This may mean putting top technical officers in operating positions so that they understand that side, or installing the top technical officer as CEO as Du Pont had done with Ed Jefferson and Merck has done with Roy Vagelos.

The CEO can understand and grapple with the key strategic questions in technology. There are really only a few of these he needs to understand well:

1. For the markets a company competes in or intends to compete in, what alternative technologies can or may meet customers' requirements? Who is pursuing each approach?
2. What are the two or three key parameters that the customer will focus on to make his buying decision? How do these key buying factors relate to the techni-

cal performance factors, or design parameters, for each technical alternative?

3. How close or far away is the company from the limits of each alternative? Said another way, how much technical potential is left? Are there ways to circumvent these limits?

4. How much will the customer value the remaining technical potential? How much will it cost to reach it?

5. What portion of the value of the technology to the customer can the producer capture given the industry structure, alternatives available to the customer, and existing or expected laws? How long will the producer be able to sustain this proportion? Will that portion be large enough and will it persist long enough to provide for adequate returns on investments? Is there any way for one producer to preempt the others in the competition?

6. When will each technology alternative be competitive in the marketplace and how long will it remain so? Can we, or anyone, develop the technology rapidly enough to meet the timing requirements?

Merely by asking and insisting on honest answers the CEO will push his company toward significantly better and sustainable strategies. And he will begin to enable it to be phoenix-like, both attacking and defending itself in an ongoing process of innovation and self-renewal.

The chief executive and chief technical officers must lead the way, and their most important task is changing the skill base of their company. But tackling the job of changing skills assumes that management has decided that the company needs to be able to cross discontinuities—that it has, in my scheme, gone beyond "strategic" management of technology which tends to allocate resources according to current growth trends. Most companies are not yet at this point. And before they can decide when and how to undertake a specific transformation of their skill base, they need to be able in general to undertake transformations. Ultimately that means a new point of view toward technological change and competition: a recognition that it is constant and unrelenting and that one doesn't fix a company's technological base once a decade but more or less continuously as new attackers appear on the horizon.

BEYOND THE LIMITS

Ruth made a big mistake when he gave
up pitching.

—Tris Speaker,
Manager, Cleveland Indians, 1921

*L*ate one night I was watching *Singin' in the Rain,* that wonderful old musical with Gene Kelly, Debbie Reynolds, and Donald O'Connor, when I realized that it was all about technological discontinuities. The plot, if you remember, focuses on two silent movie stars whose careers are threatened by a new technology—"talkies." Their friends tell them that moving pictures with sound will never sell, but soon all the Hollywood studios are trying to produce them. They try, too, but fail miserably. For one thing, the microphones pinned to their clothes make loud noises every time they move. But that's because the technology of talkies is still being worked out—struggling through the bottom of its S-curve. They circumvent this difficulty but then run into a really big problem: the glamorous female star has a high, squeaky voice. Audiences will laugh at her. So what do they do? They get Debbie Reynolds to record her lines and sing for her. A wonderful hybrid product and it works. At least for one movie. Then people find out it's really Debbie, who is just as pretty, and she becomes the star. The career of the beautiful blonde with the squeaky voice is over. Once again the leading (lady) loses.

What happened in the movie happened in real life. Many silent screen stars faded when talkies appeared. And Harry

Warner is reported to have asked, "Who the hell wants to hear actors talk?" * Why is it that we misjudge what tomorrow will bring? Even though we see new developments and use new products almost every day and as managers have at our disposal various strategic planning and forecasting models, the past remains prologue. We expect that the world will change, in fact look forward to it, but somehow we don't think that it will affect us—our jobs, our businesses.

We have seen in this book how people lose their jobs or have very difficult transitions when new technologies bring far-reaching changes to their industries. People have a tough time changing and that's why companies have a hard time changing. It's more than the initial aversion we have to new technology like automated tellers; it's the fact that we may no longer be competent in the new world that technology brings.

I don't know that I'm any better than others at accepting or embracing what the world promises to offer, but the concept of the S-curve helps me look at today and tomorrow with a pair of glasses different from most people's. Perhaps it has made me a little less vulnerable to the tendency we have to dismiss unusual events and developments as aberrations and on occasion enabled me to discern which of them represent important trends.

The past *is* prologue when people make the same mistakes and misjudgments about the future. We have looked at the past from a different angle and noted a symmetry and an inevitability about it—S-curves rising and falling, skills blossoming and fading, people succeeding and failing. When Burroughs pulled ahead of NCR in electronic cash registers, it

* Cerf, Christopher, and Navasky, Victor, *The Experts Speak,* New York: Pantheon Books, 1984.

wasn't just a case of superior product development. It was a story of hubris and a company succumbing to laws just as powerful as those that governed Greek drama. Thousands of people lost their jobs, plants were torn down, inventory destroyed and almost the entire management team fired all more swiftly than anyone would have imagined possible. But it is possible and will be even more so in the future.

Current market battles should now look different to us than they do to others. For example, most industry observers view the new fruit-juice soft drinks as mostly niche products posing little threat to the giant cola drinks. They could be wrong.

A recent Pepsi ad in *The New York Times* read, "After 87 years of going at it eyeball to eyeball, the other guy just blinked." Coke had changed its formula to more closely match Pepsi's. The uproar that followed was covered daily in all the major newspapers. But Coke's decision to match Pepsi was just one more move in the chess game that these two competitors have been playing for longer than my lifetime. That after all these years the big move should be to make the two products more the same suggests how little choice was left to either company to achieve a competitive advantage with its product or process.

In fact, leading up to the formula switch was another battle that suggested the same thing. We all remember the familiar 6½-ounce Coke bottle and its Pepsi equivalent. And we can also remember when the 6½-ounce bottle was joined on grocery store shelves and vending machines by the 12-ounce can. This advance—and some people considered it one —was so successful that it was followed with the 16-ounce glass bottle, and then the 26-ounce bottle, the 28-ounce bottle, and finally the 32-ounce bottle. From there the competition got dirty. First one company went to the 48-ounce bottle, then the other. Then both followed quickly with the 64-ounce bottle. Where would all this exciting competition lead? To the metric system when the 64-ounce bottle was replaced by the 2-liter bottle (67.4 ounces), now made of plastic, and then by the 3-liter bottle.

This bottle battle was matched by a brand battle. Coke added Diet Coke, Caffeine Free Coke, Caffeine Free Diet Coke, and Pepsi contributed Pepsi Lite, Diet Pepsi and Pepsi Free.

All this preceded the blending of Coke and Pepsi. What non-advance can we expect now? Five-liter bottles with wheels?

In fact, an advance—a real advance—may have already occurred. While the cola wars were becoming more and more marginal, Pepsi introduced Slice, a new lemon-lime–flavored juice-based soft drink. Many observers saw Slice as a competitor for 7-Up and Sprite; and indeed it was that. But it was more than that. It was the beginning of a new era of beverage technology for Pepsi. Pepsi is seeking to use a new and undeveloped technology, mass-distributed carbonated juices, to gain a competitive advantage in cold beverages. Not just to compete against 7-Up and Sprite, but to compete against Coke itself.

Juice and beverage "technology," you say. Isn't that stretching it a bit? Isn't it simply that Pepsi is trying to capitalize on the health food craze?

I don't think so. Certainly consumers are more health-conscious today than they were 20 or 25 years ago. But the changes are no different from the changes in the tire business when the consumer thought his ride on bias-ply tires had become comfortable enough and decided he wanted the better handling and longer life of radials. In both cases new technology enabled companies to capitalize on these shifts.

There is technology in juices, and in the future there will be more. For example, juices contain sugar; so if you want to create a diet-juice drink, you have to be pretty clever. You can't take the sugar out and still meet the federal requirements for what a juice is. But you can't leave it in and have a diet drink. How Pepsi has done that is a secret—a technological advantage rather than bottle or brand one-upmanship.

Then there's the question of how to preserve the juice so that it will make it through what might be months of storage and still be bright and clear and not brown the way most juices become after long storage. Juices are expensive, much more so than the ingredients in conventional soft-drinks. So processing the juices at low cost is important. The key is again a closely held technical process. In the future these costs may be dramatically affected by the new genetics (DNA) technology that is being rapidly developed. Many research centers are working to develop large fruits with higher juice content per cubic inch.

Slice is not a 100-percent-juice drink. It is a hybrid product—part juice, part traditional components—but to the consumer it is a very new offering. All the evidence so far is that the consumer loves the product. Slice has been one of the most successful non-cola product launches in the beverage industry in the last quarter century. If Nielsen is right and Slice maintains its share, it will have sales in its first year of over $400 million.

Slice could be the first entry of what may become a whole series of improved, more nutritious carbonated juice drinks. That is why I think Slice is an important innovation, not just another 7-Up. Without the S-curve and discontinuity paradigms to aid the interpretation, Slice may appear to be another attack on a niche market. But with them we are alert to the possibility that Slice may be a potent new technology that could give Pepsi a significant competitive edge in the non-cola market and perhaps someday even in the entire soft drink market. We are already seeing this happen with NutraSweet, which has begun to capture customers from sugar producers after first making inroads in the artificial sweetener market.

The perspective we have explored also adds insight to our understanding of international competition. To me the Japanese appear to have been more successful at metamorphosis, at playing the game of simultaneous attack and defense, at looking into the future a little more openly than American managers. We saw earlier how Japanese receiving-tube manufacturers made the transition early into transistors and integrated circuits. They were driven to do this by several forces and circumstances not present in the U.S.

These have since changed, but a culture that embraces aggressive change remains in many Japanese companies. A friend, a young Indian engineer who is working in a Hitachi steel foundry, tells me that next to the foundry is an advanced ceramics processing pilot plant that has been there for some time. Can steel workers really learn to run ceramics plants? The Japanese may give us the answer.

The successful transformation of the Japanese sewing machine company, Brother, is reported by Ken Ohmae.

"Brother of Japan looked at its main business, sewing machines, and realized that fewer and fewer Japanese women actually sew; they buy ready-made blouses. So Brother used its knowledge of precision machinery and microelectronics technology to diversify into office automation. It quickly became one of the world's biggest makers of electronic typewriters. Recar, another sewing machine maker, planned only around its sewing machines, in which it had a healthy market share. It concentrated on perfecting its product without adapting to changes in the market. That market was collapsing, and earlier this year Recar went out of business.

"Brother's new typewriter created a new market. It brought its sewing machine accuracy to an existing product —typewriters—and changed the world's conception of that product.

"In another example, Sumitomo Denko and Furakawa Denko, after examining their copper-wire businesses, decided to shift half to fiber optics. So despite the enormous differences between metallurgy and glass (which is used in fiber optics), both companies have diversified into optical wire." *

These cases are significant. They suggest to me that the Japanese welcome new technology. Most other countries do not. According to McKinsey's managing director in Amsterdam, Max Geldens:

"The fear that machines owned by others would destroy jobs is at least 700 years old. In 1397, the guilds in Cologne persuaded the town's government to ban a machine making heads on pins because it would cause unemployment. In sixteenth-century England, Parliament enacted a law restricting the purchase of looms with the intention of protecting handcrafted products. In the same century, Anthony Muller, an inventor of a more efficient weaving apparatus, was promptly put to death by the local mayor who feared widespread unemployment and unrest. In 1663, London workers destroyed the new, mechanical sawmills that were viewed to threaten their livelihood. Amsterdam city officials responded to the urgings of the guilds by prohibiting the construction of tower mills that could have replaced inefficient post mills

* Ohmae, Kenichi, "Transform Leader Strategies Into Golden Opportunities," *The Wall Street Journal*, December 24, 1984.

and dislocated hundreds of generously compensated millers. The tower mills were built in a different city with the effect that the Amsterdam millers were displaced anyway.

"Ribbon making machines were banned in Holland and smashed in England in 1676. In 1710, Nottingham rioters smashed stocking frames and burned warehouses full of finished goods. John Key, the inventor of the flying shuttle, was attacked by infuriated mobs who subsequently burned his house down. By 1811, machine smashers had organized themselves in gangs, called themselves Luddites (in honor of Ned Ludd who reputedly destroyed a machine to which he was assigned three decades earlier) and systematically smashed labor saving devices in all parts of England." *

Is it any different today in some European countries? Even in America the labor unions have hardly embraced automation. Japan does not seem to have these fears, or at least is covering them well for now. In a 1982 report entitled "Building an Industrial Structure for the 21st Century," the Japan Committee for Economic Development states:

"Technology has in recent years become an increasingly familiar element in our lives. Scientific and technological development that is harmonious with nature, man and society can increase social vitality and creativity.

"Today's extensive application of electronics is bringing a sweeping change in the industrial structure and the working environment. Fortunately, Japan has thus far adapted well to the qualitative changes in the labor force that have accompanied the development of electronics." †

---

* Geldens, M., "Future Employment in Industrialized Democracies," McKinsey & Company, 1984, p. 11.
† "Building an Industrial Structure for the 21st Century," Tokyo: Keizar Doyaku, 1982.

We are unlikely to see Japanese Luddites.

The S-curve and its underlying dynamics pinpoint a problem that exists in many U.S. corporations. Most of them don't do enough basic research, and if they do, it is poorly guided. In fact, probably no more than $250 million to $300 million is spent by U.S. industrial companies for basic research on their own behalf, and most of that is spent by about 10 companies, including IBM, AT&T, General Motors, Ford, General Electric, Du Pont, and Monsanto. If you talk to the CEOs of these companies, they will often wonder out loud what they have gotten for these expenditures. It's a good question, because in many instances, there have not been significant results.

But the problem may not be with the people conducting the research. They are often asked to come up with new products or new business opportunities, but researchers are not qualified to do this. To develop new products or new business one needs market and competitive insight; one needs to understand the yield side of the return equation. While researchers are failing at this task, another one, crucial if you believe the S-curve, is often left undone—determining the limits of present technologies and assessing the potential of new alternatives.

Too much science today is advanced too late, in the service of engineering. That is, research is done only as a last resort if a convenient engineering solution cannot be found. The demands of the marketplace seem to be responsible for this. For example, in the development of new chemical processes, very corrosive conditions are often encountered. These conditions can be so adverse that a new plant may not operate. Yet our understanding of the corrosion process is still weak. Correlations based on past experience are available, but these empirical observations are a limited guide to what is possible. Fundamental understanding of the science of the corrosion process is necessary to untangle and fix this problem. Even that may not be enough. The result of the investigation may tell us that we have been doing as well as we can hope to—our anticorrosive technology is at its limit. Wholly new approaches will have to be developed, perhaps with plastics.

In the pharmaceutical industry the conventional approach has been the empirical screening of compounds. Companies try hundreds of variations of chemical compounds until one combination works—molecular roulette. They don't know why it works, just that it works. Only recently have we discovered why aspirin works, decades after its "discovery." This empirical approach is slow or lucky, and getting both less productive and more expensive. Now researchers are trying to understand the chemical mechanisms of the body itself. In particular they are looking for drugs the body produces such as interferons, and trying to reproduce them.

Today the design of aircraft is based on empirical correlations between speed and drag. We need to know more about the flow of air around the plane and the onset of turbulence. We need to know about the failure mechanisms of wings and how polymers can be made more resistant to high temperature and how much more resistant. We need to know about disease mechanisms in detail so we can design drugs which intervene in a fundamentally different way than current drugs. We need to know about how amorphous (unstructured) materials work electrically so we can tell if we can use them as very cheap sources for electronic components. We need to know.

The best companies have set out a deliberate, well-focused program of internal or external acquisition of fundamental scientific knowledge. They have a strong corporate-level research activity that can grapple with tough problems that lie in the path of successful development. Time is a factor, but time is usually available because problems and limits have been anticipated. Budgets are not of overwhelming importance because they are small compared to potential sales and opportunity costs. The whole process is frequently run by a very senior scientist who has credibility and a ready communication channel to top management. This is the only way I know to ensure that research is directed toward tomorrow, toward the truly strategic needs of the business, not just current market priorities.

The long term eventually becomes the short term, and swiftly as we have seen, so understanding the S-curve might also make us better predicters of what companies' stocks will do well. One recent book * on picking high-tech stocks recommended finding answers to questions about what the company does, what its historical trends have been, the essential demand for the product, whether it is profitable or not and whether management is solid and properly motivated.

These are all standard, good questions, but they are not good enough. In my mind they give a static snapshot of the company rather than any insight about how the competitive battle of punch and counterpunch will turn out. They can easily suggest buying a stock just as its technological health begins to deteriorate but before that shows up in the financial indicators. What investors and analysts need to ask are questions that determine whether a company will be an attacker or defender or, better, both.

* Toney, A., and Tilling, T., *High Tech—How to Find and Profit From Today's New Super Stocks*, New York: Simon and Schuster, 1984. Questions cited are in the chapter "Picking Apples Instead of Lemons."

.   .   .

One final note. I find the paradigm of the S-curve and discontinuities to be positive. Opening up more possibilities than it closes down. Supporting our resolve to learn again and make our companies better learning organizations. Helping us to find and face real problems. Guiding our energies around them. If one knows when present approaches will reach their limits in the future, then one can start now, while there is time, to seek alternatives. That is what Jack Kilby and Robert Noyce did when they found ways around the tyranny of numbers. What Sir Godfrey Hounsfield did when he found a way to look through the cranial barrier into the brain. What Joe Wilson and Tom Watson did when they found ways to replace carbon paper and tabulating machines. These men and others said, know your limits *so you can go around them*. In finding ways around limits they not only provided better solutions to known problems, but opened up unanticipated opportunities on an unimagined scale.

Are You Doing
the Fundamentals?

Are your current management processes and systems likely to detect possible discontinuities? The following questions may help you make that assessment.

1. *Have you identified the key buying factors for each product group?* Understanding technology starts with understanding customers. Auto manufacturers, for example, may want something different from retailers who sell replacement tires, or at least a different mix of tread life, stability and other factors. It is surprising how few commodities there really are. Equally important is understanding how selling factors are likely to evolve and which consumers are likely to lead the way in this evolution.

2. *Is there agreement on the relationship of these buying factors to key design variables?* This requires understanding what the engineer can change in the design of the product or process: the speed of the copier, the fuel economy of the engine, the fluff of the diaper. Engineers can change just about anything if they have enough time, money and help from the research staff, but it is necessary to understand what can be changed quickly and cheaply and what will

take longer and cost more. The amount of disagreement here is surprising. Call a meeting of technical and marketing people. Beforehand, ask each group to give you their answers independently. More often than not they will be very different because communication between the technical departments and marketing is weak.

3. *What are the limits of the key design variables?* Said another way, do you know how much untapped technical potential remains in your major technologies? It is not possible to answer this without an ongoing program to seek and find limits and to determine ways around them.

4. *Have you identified your direct and indirect competitors?* Large and direct competitors are generally well known. What about smaller companies or even larger ones that are not now directly competing with you? Perhaps they are in adjacent markets to yours, or are suppliers to your competitors. For example, newly merged food and tobacco companies could move into beverages. Consumer products companies have moved into pharmaceuticals. And it was a pharmaceutical company, G. D. Searle, that came up with the best selling sugar substitute, NutraSweet.

5. *Do you know the limits of competitors' approaches?* Each of your competitors, both direct and indirect, will be trying to serve the market with a different product or different approach to delivering that product. Are your R&D units familiar with these approaches? Have they provided you with a point of view on the limits of each approach? What are the economic implications of these limits? Do your com-

petitors have alternative ways to get around their own limits?

6. *Do you know whether your R&D productivity has been rising or falling?* Few companies try to grapple with the measurement of their R&D productivity, but it is one of the best early indicators of approaching maturity. If you do not have these measurements you might look at the increases in performance and cost of development of the last several new products. Declines or increases in the rates of performance improvement divided by cost will tell you a great deal. It is important to look at this over a twenty-year period, not just five years. Things always look evolutionary over five years.

7. *Do you understand the economic consequences (i.e., impact on price and profits) of new technology introductions by you or by your competitors?* New technologies not only bring new options to purchasers, but they also bring new capacity to the marketplace. This added capacity can bring prices down no matter how much additional value is being provided to the customers. Recall the recent radical declines in the price of semiconductor memory despite the dramatically increased capability of the chips. Or the price of pharmaceuticals after they come off patent and new competitors enter the battle. Anticipating these changes before new capital is allocated can ease cash flow problems in rapidly growing businesses.

8. *Do you know which of your businesses are most vulnerable to technical attack?* A simple assessment of which business can be most affected by a wholly new technical approach (as vacuum tubes were sus-

ceptible to attack by solid-state products) is helpful for setting priorities.

9. *Do you have a plan for containing the threat?* Forethought will offset the potential for the chaos that follows a surprise attack. It also increases the effectiveness and decreases the cost of the counterattack.

10. *Is there open and frequent communication between your technical, marketing and manufacturing organizations?* This can be assessed not only by counting the number and frequency of meetings between the groups, but by examining the agendas. Given the often wide differences between the perspectives of marketers and technologists, it is possible for the two to meet, talk and still not communicate. Developing a common language based on the ideas of technical productivity and yield can help increase the effectiveness of the dialogue substantially.

11. *Is the chief technical officer part of the CEO's inner circle of confidants?* Close communication between the CEO and the chief technical officer is crucial if the corporation expects to effectively exploit its technology. One test for whether their communications are based on a common framework is whether the chief technical officer is considered as a possible candidate for the top job. If he is not, perhaps he should be replaced or moved into a line position for development.

If you have answered "no" to more than three or four of these questions, you need to explore how to improve your ability to spot and deal with technological discontinuity.

Assessing the Threat (Drawing an S-Curve)

*I*f you uncover a potential problem, you will need to become more explicit about the magnitude of the potential threat and how much effort will be required to deal with it. To do that you will need to think through the S-curves for your technology and those of your potential competitors. Four analyses are required.

1. *Identifying alternatives to your present approaches.*
    The task here is to list, not evaluate, options. For people with a common skill base it can be tough to identify options. They see the potential benefits and extensions of their own approach more easily than they can see anyone else's. Thus they discount what others might do. They implicitly confuse option generation with option evaluation.

    Paradoxically, in most corporations, somebody has a clear sense of what the alternatives are. If these people can be found, option identification can go smoothly. Often they are not sought, or if sought and found, not listened to. American companies have many technology listening posts in Japan, but I'm afraid that only a few listen to their listening posts when they report new developments.

2. *Identifying performance parameters.* The next step is to identify the performance parameters for each group of product users. As we discussed in chapter 3, each group will have a different set of needs and therefore different performance parameters. The performance parameters also must be related to key design factors of each specific technical approach. All of this is time consuming and complex. The best approach is to pick a few areas that are critical and concentrate on them rather than trying to cover the waterfront.

It's important to recognize that the performance parameters will change over time. The analysis of present and past performance parameters is usually critical for making the best guesses about what future performance parameters will be. One helpful way to get started on this analysis is to examine the past pace of change of performance parameters. Has it been fast or slow? Have the parameters changed a lot or a little? Why have they changed, i.e., because the market has changed what it wanted or because competition changed?

With these assessments in hand, one can go on to make an informed guess about future performance parameters. Will customer interests or competition change? What else will change that could affect the future parameters (e.g., laws)? When will these changes occur? Is it possible that the rates of change will be faster or slower than in the past?

3. *Calculating limits.* You need to do this for each performance parameter, which implies for each technology as well. The best way is to have groups of technical personnel with relevant background meet with your best limitists to discuss what mechanisms

might limit the performance of the technology. Will it be thermodynamics? Strength of materials? Chemistry? Laws of motion? Some fundamental physical force? Or some combination of these? This debate takes time, but the time is necessary to solve the problem. Many blind alleys will be pursued. Sometimes technical people find it easier to think about what the present limits are, and how they can be avoided, rather than going straight to the ultimate limit. If that is their preference, it's okay.

It is often impossible to say with certainty what the technology limits are until the necessary scientific understanding based on physical data is gained. Collecting the necessary data involves time and expense. Choosing the proper point at which to stop the analyses is a key decision for the technical and business people involved in the discussion.

Once the limiting mechanisms have been determined, the numerical limit can be estimated. This is best done by the technical people since it requires an understanding of the limiting mechanism and the laws of science. The remaining technical potential of the business is determined by comparing the limit with the state-of-the-art. The technical potential can then be compared with the remaining potential of other aspects of the business, at least conceptually, to determine how important technology is likely to be to the overall strategy. The potential for each technical alternative should also be compared.

It is one thing to know the ultimate limits of a technology: it is quite another to close the gap between the present state-of-the-art and that limit. To do that requires both an idea about how to close the gap and where to start. Along the way you will un-doubtedly have to experiment as you search for miss-ing information, comparing the limits of one approach to another. In this way, possible sequences of technical development can be laid out, evaluated for cost, risk and return and the best one, or ones, selected.

This is usually an appropriate stage to seek ways around the limits as well. You need to involve your best "limit breakers" in this step. Once they have a clearly defined problem before them—a defined limit —they will be able to generate alternatives to circum-vent it. Often this requires a long "soak" time. Indeed much of the best thinking may go on in the shower. Where it takes place best, however, is less important than making the task of circumventing limits explicit.

4. *Drawing S-curves.* Drawing an S-curve is a fairly straightforward process once the performance param-eters and their limits are known. The first step is to reconstruct what has happened in the past. The sec-ond step is to plot the limit. The third step is to fore-cast likely future progress.

   a. *Historical Analysis.* The first step is to reconstruct the history of product introductions. The data on the performance and cost of development of each new product is what is plotted to reveal the S-curve. For each new product brought out in the past five or ten years (depending on how quickly change is occurring), you need to collect the data on how well that product performed—its score on the performance axis. You also need to collect or estimate the amount of technical effort that went into developing the product. Usually this effort is expressed in man-years. This necessitates figuring out when the development was begun and when it ended. It also means making some decisions about what expenses to include. It doesn't really matter if you include initial marketing expenses, the cost of the first prototypes, initial sample quantities or other expenses as long as they are treated consistently.

b. *Plotting the Limit.* The second step is to express the performance limit in the same terms that you measured product performance and plot that number as a horizontal line running across the top of the graph. This establishes the top of the S-curve.

c. *Forecasting the Future.* The third step is to project the plot of historical performance into the future. There are two ways to do this. The simplest approach is to make the top part of the curve symmetrical with the bottom part and draw a straight line connecting the points. This approach is fast and cheap. It also produces the least insight.

The second approach is more mathematical. The S-curve, as your mathematician friends will tell you, is a "logistics" or "Gompertz" curve. If you know any three points on the curve you can find all the others. In fact there are computer routines that will allow you to do this on a PC. The three data points you need are two historical product introductions, the limit and a rough estimate of how much total effort it will take to reach the limit. If you are uncomfortable in venturing a guess about this number, it can be estimated by the technical staff. This approach to drawing an S-curve is more work but can provide more insight because it allows you to think about the slope of the curve and discuss how it might be improved through better management.

There is an even more sophisticated approach based on an equation by Putnam,* but it is not likely to be worth the extra time and effort.

Usually at this stage people would like to know if we actually draw S-curves. The answer is, "sometimes." S-curves are useful for understanding what has happened and for checking the reasonableness of technology plans. But constructing S-curves is time-consuming and should be done only where there is considerable debate about which course to pursue. In these cases the value of the answer received may exceed the cost of gathering the information. We do not suggest drawing S-curves as part of an annual planning process; the whole process might collapse under its own weight.

S-curves can help communications, but as a general proposition, it is more important to believe they exist, and to be able to check that belief if necessary than it is to construct them for each and every case. Often what is of most value about the S-curve is its limits, the notion of the "productivity" (or slope of the curve) and how it will change. These factors can often be estimated without going through the full-blown process of drawing each and every S-curve.

Defining the technology properly can be difficult. I have said that nylon tire-cord technology was at its limit in 1968. But since then scientists have come up with a radically restructured form of nylon that has no flat spotting so its performance has increased once again. This form of nylon was

* Project cost $= \dfrac{(\text{Projected performance})^a}{(\text{Efficiency}) \times (\text{Time})^b}$

Where a and b are constants specific to each development laboratory.

not known or even conceptualized in 1968, so it could not have been evaluated. The switch to polyester came and has lasted for fifteen years, but that is not to say that the new form of nylon will not regain share.

Bear in mind that performance parameters are, after all, defined for a particular market segment and the emergence of a new technology may change the definition of that segment itself. For example, as the home computer becomes more prevalent, it will change the way banks segment their customers. Banking performance means something different to a user of home computers than to walk-in customers.

Timing the Attack
(The Yield Analysis)

f the S-curve analysis reveals a specific threat, then it is necessary to be more explicit about when an attack might occur. Yield is the economic benefit the producer receives for advancing the technology. It is a function of the market size and the lifetime of one product in one market. If the market size and growth are known, then the key to estimating yield is to determine how long the product will last in the marketplace. Thus the following timing analyses are key to assessing yield.

1. *Conceptualize alternate products (or processes).* On the basis of an understanding of limits and performance parameters, we are now in a position to carry out a creative step in the timing analysis, which is to conceptualize competing products. More mistakes are made here than in any other area. If we cannot come to grips with what products may be competing against one another, then we will certainly never get their costs, value to the customer, and profit economics right. Moreover, estimates of when various approaches will be competitive may be far off.

   Conceptualizing these competitive products (or processes) is not easy. If one has done the research thoroughly, and if the marketing and technology departments have assembled the pertinent facts, there are a lot of alternatives to be considered. Each is likely to be a slight variant of the other. The simplest of these products needs to be plucked from the group and used as a base case; the others can be arrayed in terms of increasing complexity.

We have found it useful at this point to come up with "imaginary" products (or processes). "Imaginary" in the sense that you would never seriously try to design them in detail or attempt to introduce them to the market. Nonetheless, they help to establish ranges of the possible. They help us to understand the longest possible time the present technology can hold out and the shortest possible time until the new technology will be economic. Also, the maximum prices likely to be charged, and thus the likely maximum profit. The imaginary product may even be a guide to the minimum prices a crazed defender might charge and what that would mean for profits, or how long product transitions are likely to last and when they will start and finish. All these bounds can be discovered through the use of "imaginary" products. We can compare products that perform exactly the function existing products do but at a different cost, or products that cost exactly what present ones do, but provide different functions. Each of these comparisons can be thought through for each, or any, year into the future and thus provide a sense of the changing economics and competition over time.

What we need for all these analyses is an understanding of the perceived economic costs—perceived by customers and by competitors. We are not talking about the financial costs, which may be assembled for tax collectors or shareholders, but the key costs, the ones management uses to make decisions on when to introduce new products, when to build a new plant, when to shut down an old one. These "perceived economic costs" must include the full costs, and by this I mean opportunity costs, entry and exit costs, switching costs (for either the customer or the supplier to switch from one technical approach to another), cost of vertical integration and any other special costs imposed by government fiat or social custom. All these need to be understood, or at least addressed to get real insight.

2. *Estimate returns.* Once we have decided on what the competitive products are likely to be and what their economics are, we can estimate the returns for investing in their development. That means we need to think through the productivity (technical advance divided by effort expended) and yield (profits received for the advanced technology). The key to the productivity analysis is thinking through the S-curve given what is known about past progress and effort expended to reach the limits of a technology. Since results will be affected by knowledge building, timing and aggressiveness of targets, computer support, and communications, these things need to be thought through as well.

It's important here to bear in mind that yield is affected by demand and supply levels, value of the product in use to the customer, industry structure, and the collective strategies of all the competitors. Even though a product may bring a lot of value to the customer, there is no assurance that any of that value will accrue to the manufacturer. If the industry is overcrowded with competitors (as most high technology industries are today), it is unlikely that any producer, be it an innovator or second to market, will make sufficient returns.

Given these two analyses as well as the S-curve described in Appendix 2, you should be able to take a guess at the timing of competitive technologies. We have found it most helpful here to project the costs of the defending and attacking products on the chart. The vertical axis of the chart plots cost and the horizontal axis time.

The first step is to plot estimates of the future (say five–ten years) variable manufacturing costs of the defending product. These estimates can best be supplied by your engineering staff.

The second step is to plot the projected full costs (variable costs plus depreciation, interest charges and return on capital) over time. The difference between these two plots is the contribution margin of the defending product in the absence of a threat from a new product, assuming that over time the price will approximate the full cost (a good approximation for most businesses).

The third step is to plot the projected full and variable costs for the attacking product. This step requires the inputs described above on "imaginary" products. This curve will cross the full and variable cost level at four points, each of which has special significance (Exhibit 26).

1. The first point in time where an economic challenge can be made to the defender. Where the full cost of the attacking product is equal to the full cost of the defending product.

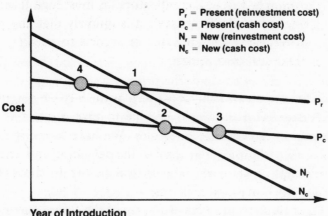

TRANSITION PRICES

P_r = Present (reinvestment cost)
P_c = Present (cash cost)
N_r = New (reinvestment cost)
N_c = New (cash cost)

Cost

Year of Introduction

26 Transition Economics.
A plot of the projected full and variable costs for an attacking and defending product. The four points describe the basic timing of a transition: when it will begin and how long it will probably last.

2. The latest point in time one would expect an attack to be launched. Where the variable cash cost of the attacking product is equal to the variable cash cost of the defending product.
3. The first point that one would expect the attack to be over. Where the variable cash cost of the attacking product is equal to the variable cash cost of the defending product.
4. The latest point in time one would expect the attack to end. Where the full cost of the attacking product is equal to the variable cash cost of the defending product.

These four points describe the basic timing of the transition—when it will begin and how long it probably will last.

They also provide a context for estimating prices. Before the transition begins prices will tend to follow the full cost of the defending product. After the transition begins they will begin to track the full cost of the attacking product, since that is the lower cost option, unless there is excess capacity. If there is excess capacity the prices can drop as low as the variable cash cost of the defending product until the transition is over. At that point they can fall as low as the variable cash cost of the attacker if there is still overcapacity. If overcapacity does not exist at that point, prices will tend to follow the full cost of the attacking technology. With these perspectives on prices and a knowledge of costs, one can calculate profitability. When the total returns to the manufacturers of the competing technologies are equal, the rapid collapse described in chapter 7 is probably near. The transition will be over and yield will go to zero.

The timing analyses should enable a company to take a proactive posture to the marketplace. They should build confidence. But they can seem precise and be wrong, too. Not because the calculations are wrong but because of the inherent inability any of us have to predict what the customer, or the competition, will do. The potential for error does not justify their rejection, however. All analysis has this disadvantage. Very few sciences are exact, and economics is not one of them. The tests are worthwhile, though, because they can help to outline the problem.

BIBLIOGRAPHY

Abernathy, Clark, and Kantrow, Alan. *Industrial Renaissance.* New York, NY: Basic Books, 1983.
Abernathy, William J. *The Productivity Dilemma.* Baltimore, MD: Johns Hopkins University Press, 1978.
American Academy of Arts & Sciences. *American Academy of Arts & Sciences Proceedings,* "Daedalus." Cambridge, MA: 1974.
Andrews, Frank, ed. *Scientific Productivity.* New York, NY: Cambridge University Press, 1979.
Angelucci, Enzo, and Cucari, Attilio. *Ships.* New York, NY: McGraw-Hill, 1977.
Archer, L. Bruce. *Technological Innovation—A Methodology.* London, Peridon, Ltd., Inforlink, Science Policy Foundation, 1971.
Augarten, Stan. *State of the Art.* New Haven, CT: Ticknor & Fields, 1983.
Bacon, J., and Brown, J. K. *Board of Directors, Perspectives & Practices.* New York, NY: The Conference Board, 1977.
Balthasar, H., Boschi, R., and Menke, M. "Calling the Shots in R&D," *Harvard Business Review,* May/June, 1978.
Bartlett, John. *Familiar Quotations.* Boston, MA: Little Brown & Company, 1980.
Basche, James R., Jr. *Production Cost Trends and Outlook.* New York, NY: The Conference Board, 1977.
Battelle. *Interaction of Science & Technology in Innovative Process.* Columbus, OH: Battelle for National Science Foundation, 1973.

————. *Probable Levels of R&D Expenditures in 1977*. Columbus, OH: Battelle Memorial Institute, 1976.

Bell Laboratories. "Record." Murray Hill, NJ: January 1975.

Bernstein, Jeremy. *Three Degrees Above Zero*. New York, NY: Charles Scribner's Sons, 1984.

Bisio, A., and Gastwirt, L. "Turning R&D Into Profits, A Systematic Approach." New York, NY: Amacom, American Management Association, 1979.

Blackman, A. Wade. *United Aircraft Research Laboratories, UAR-H285*. Cambridge, MA: An Industrial Dynamics Model of UARL Growth, ABT Association, 1969.

Bode, H. W. *Synergy: Technical Integration & Innovation in Bell System*. Murray Hill, NJ: Bell Laboratories, 1972.

Bozeman, Crow & Link. *Strategic Management of Industrial R&D*. Lexington, MA: Lexington Books, D. C. Heath, 1984.

Brancazio, Peter J. *Sportscience*. New York, NY: Simon and Schuster, 1984.

Bright, James R. *Practical Technology Forecasting Concepts & Exercises*. Austin, TX: Industrial Management Center, 1978.

Brooks, Frederick. *The Mythical Man-Month*. Reading, MA: Addison-Wesley Publishing, 1982.

Brown, J., and Elvers, L., eds. "R&D: Key Issues for Management." New York, NY: The Conference Board, 1983.

Brown, J., and O'Connor, R. "Planning and the Corporate Planning Director." New York, NY: The Conference Board, 1974.

"Building an Industrial Structure for the 21st Century," Keizar Doyaku, Tokyo, 1982.

Business Week. February 11, 1985.

————. "A Research Wiz Steps Up From a Lab," June 24, 1985.

————. "Harris Is Raising Its Bet on the Office of the Future," July 18, 1983.

————. "How Chesebrough-Pond's Put Nail Polish in a Pen," October 8, 1984.

————. "The Video Market Japan May Snap," March 1, 1982.

————. "There's a New Team at Motorola," April 23, 1966.

Bussey, John. *The Wall Street Journal*, October 3, 1984.

Center for History of Chemistry, University of Pennsylvania, *CHOC News*, Spring 1983, Vol. 1, No. 2.

Cerf, Christopher, and Navasky, Victor. *The Experts Speak*. New York, NY: Pantheon Books, 1984.

Chapelle, H.I. *The Search for Speed Under Sail*. New York, NY: W.W. Norton, 1967.

Clifford, Donald K., and Cavanagh, Richard E. *The Winning Performance*. New York, NY: Bantam, 1985.

Comm. for Economic Development. "Stimulating Technological Progress." New York, NY: Comm. for Economic Development, R&D Policy Comm., 1980.

Comm. on Science & Astronautics (Hearings). "Science, Technology & the Economy." Washington, DC: U.S. Government Printing Office, 1971.

Constant, Edward W. "The Origins of the Turbojet Revolution." Baltimore, MD: Johns Hopkins University, 1980.

Creamer, Daniel. "Overseas Research & Development by U.S. Multinationals 1966–75." New York, NY: The Conference Board, 1976.

De Solla Price, Derek J. *Little Science, Big Science*. New York, NY: Columbia University Press, 1963.

Dearborn, D., Kneznek, R., and Anthony, R. "Spending for Inds. Research 1951–52." Boston, MA: Harvard University Division of Research, School of Business Administration, 1953.

Dertouzos, Michael, and Moses, Joel, eds. *The Computer Age, 20-Year Review*. Cambridge, MA: M.I.T. Press, 1980.

Dickson, D. "Technology and Cycles of Boom and Bust," *Science*, February 25, 1983.

Dreman. *The Contrarian Strategy of Investment*. New York, NY: Random House, 1979.

Drucker, Peter. *The Age of Discontinuity*. New York, NY: Harper & Row, 1969.

———. *Innovation and Entrepreneurship*. New York, NY: Harper & Row, 1985.

———. *Landmarks of Tomorrow*. New York, NY: Harper & Row, 1959.

———. *Managing in Turbulent Times*. New York, NY: Harper & Row, 1980.

Duerr, M., and Roach, J. "Organization and Control of International Operations." New York, NY: The Conference Board, 1973.

The Economist. "The New Entrepreneurs," December 24, 1983.
Electronics, April 17, 1980.
European Industrial Resource Management Assoc. "Methods for the Evaluation of R&D Projects." Paris, Eirma. 1970.
———. "Planning for Research & Development." Paris, Eirma (Speeches, Rotterdam), 1977.
Fairbain, William. Merchant Sail. *The Economist.* "The New Entrepreneurs," December 24, 1983.
Fenn, John B. *Engines, Energy and Entropy.* San Francisco, CA: W.H. Freeman, 1982.
Flaherty, R.I. "Harris Corporation's Remarkable Metamorphosis," *Forbes,* May 26, 1980.
Forbes. September 15, 1977.
———. "Winning With Wang," October 15, 1976.
Ford, Henry, with Crowther, Samuel. *My Life and Work.* Garden City, NY: Doubleday & Company, 1922.
Foster, Richard N. "A Call for Vision in Managing Technology," *Business Week,* May 24, 1982.
———. "Improving the Return on R&D," *Research Management,* January/February, March/April, 1985.
———. "Timing Technological Transitions," *Technology and Society,* Vol. 7, Nos. 2 and 3, 1986.
———. "To Exploit New Technology, Know When to Junk the Old," *The Wall Street Journal,* May 2, 1983.
———. "Why America's Technology Leaders Tend to Lose," *Vital Speeches,* June 15, 1983.
Fortune. "The Michelin Man Rolls Onto Akron's Backyard," December 1974.
Freiberger, Paul and Swaine, Michael. *Fire in the Valley: The Making of the Personal Computer.* Berkeley, CA: Osborne/McGraw-Hill, 1984.
Friedel, Robert. *Pioneer Plastic, Making & Selling of Celluloid.* Madison, WI: University of Wisconsin Press, 1983.
Fruhan, William E., Jr. *Financial Strategy.* Homewood, IL: Richard D. Irwin, 1979.
Fusfeld, H., and Langlois, R., eds. *Understanding R&D Productivity.* New York, NY: Pergamon Press, 1982.
Gardner, Howard. *Frames of Mind.* New York, NY: Basic Books, 1983.

Gardner, John W. *Self-Renewal*. New York, NY: W. W. Norton, 1981.

Gardner, M. "The Computer as Scientist," *Discover*, June 1985.

Geldens, M. *Future Employment in Industrialized Democracies*. McKinsey & Company, 1984.

Gellein, Oscar and Newman, Maurice. *Accounting for Research & Development Expenses*. New York, NY: American Institute of CPAs, 1973.

Gilfillan, S. C. *The Sociology of Invention*. Cambridge, MA: M.I.T. Press, 1970.

Goodenough, Simon. *Tactical Genius in Battle*. New York, NY: E. P. Dutton, 1979.

Government Research Corporation. *National Journal*. Washington, DC, March 15, 1980.

Grabowski, Henry G. *Drug Regulation and Innovation*. Washington, DC: Enterprise Institute for Public Policy Research, 1976.

Grabowski, H., and Vernon, J. *Impact of Regulation on Industrial Innovation*. Washington, DC: National Academy of Sciences, 1979.

Gray, Irwin. *The Engineer in Transition to Management*. New York, NY: IEEE Press, 1979.

Gutting, Gary, ed. *Paradigms & Revolutions*. Notre Dame, IN: University of Notre Dame Press, 1980.

Hanna, Ali. "Teaming Up With Technology." Speech, Copper Development, Inc. Hilton Head, SC: 1984 Annual Spring Meeting, May 7–8, 1984.

Hart, B. H. Liddell. *Signet*. "Strategy." New York, NY: New American Library, 1974.

Harth, Erich. *Windows on the Mind*. New York, NY: William Morrow, 1982.

Havenmeyer, Loomis, and Dudley, Samuel. *Engineering Heritage at Yale*. New Haven, CT: Yale University, 1959.

Heller, Robert. *The Super Managers*. New York, NY: E. P. Dutton, 1984.

Hirschmann, Winfred. "Profit from the Learning Curve," *Harvard Business Review*, January/February 1964.

Holton, G., and Morison, R., eds. *Limits of Scientific Inquiry*. New York, NY: W. W. Norton, 1979.

Hopkins, David. *New-Product Winners and Losers.* New York, NY: The Conference Board, 1980.

Houghton, J. R. "The Role of Technology in Restructuring a Company," *Research Management,* November/December 1983.

Hudson, R., Chambers, J., and Johnston, R. "New Product Planning Decisions Under Uncertainty," *Interfaces,* November 1977.

Hughes Aircraft. *R&D Productivity,* 2nd ed. Culver City, CA: Hughes Aircraft, 1978.

IIT Research Inst. *Technology in Retrospect & Critical Events in Science.* Washington, DC: National Science Foundation, 1969.

Illinois Institute of Technology. *Technology Retrospect & Critical Events in Science.* Washington, DC: National Science Foundation, 1968.

Indicators of Science and Technology. Tokyo. 1979.

Industrial Research Institute. Reprinted from *Research Management.* "Chief Executive Officers Speak Out on Industry."

Jacob, Francois. *The Possible & the Actual.* New York, NY: Pantheon Books, 1982.

Janger, Allen R. *Organization of International Joint Ventures.* New York, NY: The Conference Board, 1980.

Jankow, R. K. "The Total Artificial Heart," *Scientific American,* January 1981.

Japan Committee for Economic Development. *Building on Industrial Structure for 21st Century.* Tokyo. 1982.

Jarvik, R. K. "The Total Artificial Heart," *Scientific American,* January 1981.

Jewkes, J., Sawers, D., and Stillerman, R. *Sources of Invention.* New York, NY: W.W. Norton, 1969.

Judson, Horace F. *The Search for Solutions.* New York, NY: Holt, Rinehart & Winston, 1980.

Kantor, Rosabeth Moss. *The Change Masters.* New York, NY: Simon and Schuster, 1983.

Kantrow, Alan, ed. "Survival Strategies for American Industry," *Harvard Business Review.* New York, NY: John Wiley & Sons, 1983.

Kantrow, Alan, et al. *Industrial Renaissance.* New York, NY: Basic Books, 1983.

Kidder, Tracy. *The Soul of a New Machine.* Boston, MA: Little Brown & Company, 1981.

Klein, Burton H. *Dynamic Economics.* Cambridge, MA: Harvard University Press, 1977.

Knight, Geoffrey. *Concorde, the Inside Story.* New York, NY: Stein & Day, 1976.

Kuhn, Thomas S. *The Structure of Scientific Revolutions.* Chicago, IL: University of Chicago Press, 1970.

Lakatos, Imre, and Musgrave, Alan, eds. *Criticism & the Growth of Knowledge.* New York, NY: Cambridge University Press, 1979.

Landes, David S. *Revolution in Time.* Cambridge, MA: Belknap Press, Harvard University, 1983.

Lawrence, Robert Z. *Can America Compete?* Washington, DC: Brookings Institute, 1984.

Levitt, Theodore. "Marketing Myopia," *Harvard Business Review,* September/October 1975.

Little, A. D. *Barriers to Innovation in Industry.* Washington, DC: National Science Foundation, September 1973.

Lorie, James, and Hamilton, Mary. *The Stock Market—Theories & Evidence.* Homewood, IL: Richard D. Irwin, 1973.

MacMillan, Ian C. *Strategy Formulation: Political Concepts.* St. Paul, MN: West Publishing, 1978.

Maddock, Sir Ieuan. "Why Industry Must Learn to Forget," *New Scientist,* February 11, 1982.

Maidique, M. A. "Entrepreneurs, Champions, and Technological Innovation," *Sloan Management Review,* Winter 1980.

Malik, Rex. *And Tomorrow the World? Inside IBM.* London: Millington Ltd., 1975.

Management Today. "Gould's Golden Gamble." February 1984.

Margenau, Henry, and Bergamini, David. *The Scientist.* New York, NY: Time Inc., 1964.

Marshall & Meckling. *Predictability of the Costs, Time and Success of Development.* Santa Monica, CA: Rand Corporation, December 1959.

Martino, Joseph P. *The Futurist.* "Science Indicators: Charting the Progress of Research." February 1975.

Matsuoka, Toshio, ed. *Japan 1980, An International Comparison.* Tokyo: Keizai Koho Center, 1980.

McClellan, Stephen. *The Coming Computer Industry Shakeout.* New York, NY: John Wiley, 1984.

McCloy, John J. *The Great Oil Spill.* New York, NY: Chelsea House Publishers, 1976.

McCorduck, Pamela. *Machines Who Think.* San Francisco, CA: W. H. Freeman, 1979.

McKinsey & Company. *Challenges Facing Insurance Executives.* New York, NY: McKinsey & Company, 1977.

————. *Strategic Leadership: Challenge to Chairmen.* London, England: McKinsey & Company, 1978.

————. *Technology Seminar.* New York, NY: Harvard Club, February 1974.

Melegari, Vezio. *Great Military Sieges.* New York, NY: Exeter Books, 1972.

Mencher, Alan, ed. *Management and Technology, Vol. 1.* Elmsford, NY: Pergamon Press for Inforlink, Science Policy Foundation, 1972.

Mensch, Gerhard. *Stalemate in Technology.* Cambridge, MA: Ballinger Publishing, 1979.

Merrow, Chapel, and Worthing. *Revisions of Cost Estimation in New Technologies.* Santa Monica, CA: Rand Corporation, July 1979.

Merrow, E. W. et al. "Understanding Cost Growth and Performance." Shortfalls in Pioneer Process Plants. Rand, 1981.

Midwest Research Institute. *Economic Impact of Stimulated Technological Actions.* Washington, DC: National Aeronautics & Space Admin., Final Report, 1971.

Miller, G. William, ed. *Regrowing the American Economy.* Englewood Cliffs, NJ: Prentice-Hall, 1983.

Miller, Lawrence. *American Spirit.* New York, NY: William Morrow, 1984.

Moffitt, Phillip. "The Dark Side of Excellence," *Esquire,* December 1985.

Morgan, C., and Langford, D. *Facts and Fallacies.* Exeter, England: Webb & Bower, 1981.

Morison, Elting E. *Men, Machines and Modern Times.* Cambridge, MA: M.I.T. Press, 1966.

Morris, B. "Beatrice Co. Says Presidents of Two Units Quit," *The Wall Street Journal,* March 1985.

Nason, H., Steger, J., and Manners, G. *Support of Basic Research by Industry.* Washington, DC: National Science Foundation, 1978.

National Academy of Sciences. *Frontiers in Science & Technology.* New York, NY: W. H. Freeman, 1983.

———. *Issues in Science and Technology.* Washington, DC. Fall 1984.

———. *Science and Technology, A 5-Year Outlook.* San Francisco, CA: W. H. Freeman, 1979.

National Research Council. *Outlook for Science & Technology.* San Francisco, CA: W. H. Freeman, 1982.

———. *Material Advisory Board, Principles of Research-Engineering Interaction.* Washington, DC: National Academy of Sciences, 1966.

National Science Board Report. *Science Indicators, 1974.* Washington, DC: National Science Foundation, 1975.

———. *Science Indicators, 1978.* Washington, DC: National Science Foundation, 1979.

———. *Science Indicators, 1982.* Washington, DC: National Science Foundation, 1983.

National Science Foundation. *International Science & Technical Data Update.* Washington, DC: National Science Foundation, January 1984.

———. *Methodological Aspects of Statistics on R&D Costs and Manpower.* Washington, DC: American Statistical Association Meeting. December 1958.

The New York Times. "Wang Labs Head to Head Against IBM," February 24, 1980.

Norman, Colin. *Knowledge and Power, the Global Research.* Washington, DC: Worldwatch Institute, 1979.

North-Holland Publishing. "The Netherlands," *Research Policy Journal,* November 1974.

Norton-Taylor, Duncan, ed. *For Some the Dream Came True; 50 Years of Fortune.* Secaucus, NJ: Lyle Stuart, 1981.

O'Connor, Rochelle. *Corporate Guides to Long-Range Planning.* New York, NY: The Conference Board, 1976.

————. *Planning Under Uncertainty.* New York, NY: The Conference Board, 1978.

O'Neill, Gerald K. *The Technology Edge.* New York, NY: Simon and Schuster, 1983.

OECD. *Gaps in Technology Between Member Countries.* Third Ministerial Meeting on Science of OECD, Paris, France, 1970.

————. *Group on New Concepts of Science Policy; Science Growth & Society.* Paris, France, 1971.

————. *Science & Technology Indicators, Resources Devoted to R&D.* Paris, France, 1984.

Office of Foreign Secretary. *U.S. International Firms & Research, Development, & Engineering in Developing Countries.* Washington, DC: National Academy of Sciences, 1973.

Office of Technology Assessment, U.S. Government. *Patent Profiles—Microelectronics.* Washington, DC: Dept. of Commerce, Office of Technology, 1981.

————. *Industrial Robots (69–3/82).* Washington, DC: Dept. of Commerce, Natl. Tech. Info. Svc., 1982.

————. *Patent Profiles—Biotechnology.* Washington, DC: Department of Commerce, Patent & Trademark Office, 1983.

————. *Patent Profiles—Microelectronics.* Washington, DC: Dept. of Commerce, Natl. Tech. Info. Svc., 1983.

Office of Technology Assess. & Forecast. *Patent Profiles—Synthetic Fuels.* Washington, DC: Dept. of Commerce, Patent & Trademark Office, 1979.

————. *Technological Innovation and Regional Economic Development.* Washington, DC. July 1984.

Ohmae, Kenichi. "Key Factors for Survival in the Microelectronics Industry." Address at National American Society for Corporate Planning. May 1984.

————. "Transform Leader Strategies into Golden Opportunities," *The Wall Street Journal,* December 24, 1984.

————. *Triad Power.* New York, NY: The Free Press, 1985.

Ord-Hume, Arthur, W. J. G. *Perpetual Motion.* New York, NY: St. Martin's Press, 1980.

Osborne, Adam. "Running Wild," *The Economist,* December 24, 1983.

O'Toole, James. *Vanguard Management.* New York, NY: Doubleday & Co., 1985.

Pascarella, Perry. "Are You Investing in the Wrong Technology?" *Industry Week,* July 25, 1983.

————. *Technology, Fire in a Dark World.* New York, NY: Van Nostrand Reinhold Co., 1979.

Peat, Marwick Mitchell. "North Sea Costs Escalation Study." London: Her Majesty's Stationery Office, 1975.

Perpich, Joseph, ed. *Biotechnology, Technology in Society Journal.* New York, NY: Pergamon Press, 1984.

Peters, Thomas J., and Waterman, Robert H. *In Search of Excellence.* New York, NY: Harper & Row, 1982.

Porter, Michael. *Competitive Advantage.* New York, NY: The Free Press, 1985.

Pry, Robert. "Organizing for Technology Implementation." Speech in Orlando, FL. November 1984.

Putnam, Lawrence. "A General Empirical Solution to Macro Software Sizing and Estimating Problems," *EEE Transactions of Software Eng.* July 1978.

Putnam, L., and Fitzsimmons, A. "Estimating Software Costs," *Datamation,* September 1979. Continued in October and November 1979.

Quinn, James Bryan, "Large Scale Innovation: Managing Chaos," *Tuck Today,* June 1985.

————. *Strategies for Change, Logical Incrementalism.* Homewood, IL.: Richard D. Irwin, 1980.

Reed, T. R. *The Chip: The Microelectronics Revolution & the Men Who Made It.* New York, NY: Simon and Schuster, 1985.

Roberts, David. "Citizen Jones," *Ultrasport,* September/October 1985.

Rogers, E., and Larsen, J. *Silicon Valley Fever.* New York, NY: Basic Books, 1984.

Rosegger, G. *Adjustments of Domestic Capacity & Production to Technology Based Imported Competition.* Amsterdam: Elsevier, 1982.

Rosenau, Milton D. *Innovation.* Belmont, CA: Lifetime Learning Publications, 1982.

Rosenbloom, R. *Research on Technological Innovation, Management and Policy.* Greenwich, CT: Jai Press, 1983.

Rushton, Gerald. *Echoes of the Whistle.* Vancouver: Douglas & McIntyre, 1980.

Sahal, Devenda. *Patterns of Technological Change*. Reading, MA: Addison-Wesley Publishing, 1981.

Schigall, Oscar. *Eyes on Tomorrow*. Chicago, IL: J. G. Ferguson Publishing, 1981.

Schmeck, H. M., Jr. "Protein That Lets Off Growth If a Human Organ Is Found," *The New York Times*, September 17, 1985.

Simon, Herbert A. *The New Science of Management Decision*. Englewood Cliffs, NJ: Prentice-Hall, 1977.

———. *The Sciences of the Artificial*. Cambridge, MA: M.I.T. Press, 1984.

Society of Research Administrators. *Journal*. Corona Del Mar, CA., 1976.

Springer, S., and Deutsch, G. *Left Brain, Right Brain*. San Francisco, CA: W. H. Freeman, 1981.

Summers, Robert. *Costs Est. as Predictors of Actual Weapon Costs*. Santa Monica, CA: Rand Corporation. March 1965.

Tanenbaum, Morris. "Relationships Between Scientific Research and Industrial Innovation." Speech at 2nd Symposium. Development & Material Technology. Israel. 1972.

Taylor, Lisa, ed. *The Phenomenon of Change*. New York, NY: Smithsonian Institute, 1984.

Thomas, Lewis. *Lives of a Cell*. New York, NY: Penguin Books, 1978.

———. *The Youngest Science*. New York, NY: The Viking Press, 1983.

Tilton, John E. *International Diffusion of Technology*. Washington, DC: Brookings Institute, 1971.

Toney and Tilling. *High Tech*. New York, NY: Simon and Schuster, 1984.

Trefil, James. *A Scientist and the Seashore*. New York, NY: Charles Scribner's Sons, 1984.

Tributsch, Helmut. *How Life Learned to Live*. Cambridge, MA: M.I.T. Press, 1982.

Trimm, D. L. *Design of Industrial Catalysts*. Amsterdam: Elsevier Scientific Publishing, 1980.

Turnbull, Stephen. *The Book of the Samurai*. New York, NY: Gallery Books, 1982.

Uttal, Bro. "Is IBM Playing Too Tough?" *Fortune,* December 10, 1984.

The Wall Street Journal. "Transforming Leader Strategies Into Golden Opportunities," December 24, 1984.

Wise, T. A. "IBM's $5,000,000,000 Gamble," *Fortune,* September 1966.

Worrall, J., and Currie, G. *Imre Lakatos Philosophical Papers,* Vol. 1. New York, NY: Cambridge University Press, 1978.

Yip, Gregory S. *Barriers to Entry.* Lexington, MA: Lexington Books, D. C. Heath, 1982.

Young, Peter, ed. *Great Generals and Their Battles.* New York, NY: The Military Press, 1984.

INDEX

technological change difficult for, 139–43
technological health of, 153–54
see also attackers; leaders
Defense Department, U.S., 231
Defiance, 65
Delta, 189
Denver Broncos, 155
detergents, *see* laundry detergents
DeVries, William, 89, 96
Dexcil, 225
diapers, disposable, 36, 38, 84, 155, 227–28
digital switches, 104
discontinuities, 35–37, 47–58
  anticipation of, 40–43, 66, 153
  in chemical industry, 116–21
  corporate death caused by, 115–135
  crossing of, 35–37, 103, 235–237, 245–47
  customers' wants and, 154–56
  defined, 35–37, 102–3
  detecting of, 214–18
  duration of, 182
  in electronics industry, 71–74, 129–30, 132–34, 135, 141–143
  frequency of, 36, 37, 47–51
  hindsight and, 235–37
  leaders affected by, 115–35
  management of, 35–37, 57–58, 101–6, 146, 149, 223–37, 245–47
  in niche vs. main market sector, 155
  R&D multipliers at, 179–82
  skill transitions and, 146, 205–210
  speed of transitions and, 161–164, 210–12
  in sugar industry, 130–31
  in tire industry, 121–28, 134–135

waves of, 51–53
see also limits; S-curves
Displaywriter, 201
Dively, George S., 233, 242
Dodd, Chester, 134
Douglas Aircraft, 187
Dow Chemical, 231
Dow-Corning, 204, 230–31
Drucker, Peter, 48, 165
drug industry, *see* pharmaceutical industry
Du Pont, 50, 56, 164
  internal organization of, 125–127
  and S-curves for tire cord materials, 123–26
Dutt, James L., 226

Eastman, George, 242
Eastman Kodak, 116, 156–57, 201
Edison, Thomas, 230, 242
efficiency, 41
  effectiveness vs., 106–11, 211
Eimac, 134
Eisenhower, Dwight, D., 107
electricity-generating machine, 80–81
electron-beam lithography, 72
electronic logic, 71
electronics industry, 38, 157, 192
  discontinuities in, 71–74, 129–130, 132–34, 135, 141–43
  in Japan, 147–49
  limits in, 71–76, 141–43
  market-share transitions in, 162
  pace of innovation in, 52
  productivity advantages in, 108–109, 170
  technological transitions in, 133, 141–43
  see also chips; transistors
EMI Ltd., 193–94
energy band gaps, 129

zero or negative, 172–73
random-access memory, 108–9
Rattler, 198
Rawland, 134
rayon, 122–23, 151
RCA, 132–34, 141–43, 201
Recar, 256–57
recorded music industry, 36, 38,
     103
Reed, John, 49, 242
Reid, T. R., 35, 82*n*
research, basic, 105–6
research and development, *see*
     R&D
Research Triangle Institute, 196
resource allocation, efficiency vs.
     effectiveness in, 107, 110–11
Reynolds Metals, 115
ribbon machines, 230–31, 257–58
risk:
   diversification and, 204–5
   in innovation, 30
   technological discontinuities
      and, 36
risk-reward ratio, 205
Roberts, David, 49*n*
Robotic Vision Systems, 51
Rockwell International
     Corporation, 143
Rommel, Erwin, 107
Rose, Pete, 236
Ross, Ian, 71–75, 76, 77, 79
Royal Society of England, 164
ruboer, synthetic, 53
runners, 61–62

sail/steamship hybrid, 197, 198
Sanyo, 148
satellite communications, 76
Schigall, Oscar, 228–29, 235–36
Schlumberger, 225
Schmeck, H. M., Jr., 150*n*
Schneiderman, Howard, 150, 207,
     210

Schroeder, William, 96
Schumpeter, Joseph, 52
Scientific American, 82
S-curves, 22, 31–32, 87–111, 122
   adding sails on, 193, 196–97
   for artificial heart research, 89,
      95, 96–97, 98
   changing performance
      parameters and, 97–101
   discontinuity as gap between, 35–
      37, 102–3
   drawing of, 269–77
   efficiency vs. effectiveness and,
      106–11
   effort vs. time factor in, 100–
      101
   forecasting with, 99–101
   leapfrogging up, 193–95, 213
   in low-tech industries, 130–31
   management's analysis of, 41,
      43, 103
   naming of, 35
   for naphthalene, 120–21
   in pairs, 102–3
   predictability of, 66
   sinewy shape of, 98
   slope of, 101, 106, 124
   technological productivity along,
      101–2, 110
   for tire cord, 121–28
   see also discontinuities; limits
Search for Speed Under Sail, The
     (Chapelle), 65
seed business, biotechnology and,
     50
semiconductors, *see* chips;
     electronics industry
Seneker, Harold, 140*n*
shipbuilding industry, 27, 28, 34,
     43, 67
   hybrids in, 197, 198
   technological limits in, 65–
      66
Siemens, 194